CONTENTS

Preface   vii

Acknowledgments   x

Using the Diskette   xi

1   Introduction to Excel   1

   1.1   What Is a Spreadsheet? / 1

   1.2   Fundamental Spreadsheet Operations / 2

      1.2.1   The Spreadsheet and the Excel Workspace / 2

      1.2.2   Excel Commands / 4

      1.2.3   Moving around the Spreadsheet / 7

      1.2.4   Cell References / 8

      1.2.5   Ranges and Arrays / 9

      1.2.6   Entering Data / 10

      1.2.7   Editing Data / 12

   1.3   Mathematical Operations on Cells—Entering Formulas / 14

      1.3.1   Formulas / 14

6.4   Spectral Data Processing / 154

6.5   Digital Averaging Techniques / 155

    6.5.1   Moving Average Filter / 156

    6.5.2   The Savitzky–Golay Filter / 157

    6.5.3   Suggested Additional Activities / 161

6.6   Feature Removal through Integration / 162

6.7   Feature Enhancement through Differentiation / 163

6.8   Conclusion / 169

    References / 171

7   Processing Experimental Data Using Solver   **172**

7.1   Introduction / 172

7.2   Using *Solver* / 173

7.3   Case Study: Deconvoluting Two Overlapping Gaussian Peaks / 181

7.4   Fitting Experimental Data / 183

7.5   Case Studies Involving Experimental Data / 183

    7.5.1   Modeling Chromatography Peaks / 183

    7.5.2   Modeling Fluorescence Decay Processes / 190

    7.5.3   Modeling of Ion-Selective Electrode (ISE) Dynamic
    Response in Flow-Injection Analysis / 197

    7.5.4   Modeling the Nikolskii–Eisenman Equation / 201

    7.5.5   Determining Rate Constants for a Ligand
    Replacement Reaction / 203

    7.5.6   Michaelis–Menten Enzyme Kinetics / 207

7.6   Conclusion / 212

    References / 213

8   Case Studies Involving VBA   **214**

8.1   Titration of a Strong Acid with a Strong Base / 214

8.2   Buffer Solutions / 226

8.3   The Marcus Treatment of Electrochemical Kinetics / 231

    8.3.1   Surface Immobilized Redox Couples / 231

Appendix A   **237**

Index   **241**

PREFACE

SPREADSHEETS AND SCIENTIFIC DATA PROCESSING

With the rapid improvements in the price–performance index in computers, the merging of telecommunications, computing, and information management, and the development of powerful portable computers, literacy in the use of this technology will become imperative for practicing scientists. Scientists these days need to be much more multifunctional than before and need to be literate in the use of computers for purposes such as graphing, processing, and transforming experimental data, investigating mathematical models, preparing reports, documents, and presentations, and for communications. Spreadsheets have a vital role to play in the development of this literacy as they enable individuals to explore data relationships in an individual and customized manner. Repetitive data processing tasks can be easily automated using macros and user-friendly interfaces easily developed with VBA routines. The ease of transfer of data and graphics between packages makes the preparation of integrated documents and high-quality graphical presentations very simple. Such presentations can also be almost instantly transferred to remote locations in electronic form using E-MAIL.

The use of spreadsheets for teaching scientific mathematics and for processing experimental data is only beginning to emerge, but clearly, it will develop rapidly in the coming years. Excel provides a marvelous environment for the development of teaching material that will be required for the growth of this literacy. We hope that this book will help provide some basic material for teachers, researchers, and scientists who wish to get involved in spreadsheet applications.

While the material is primarily related to chemistry, the approach is generic and can be applied to many situations in physics, engineering, and biology.

In preparing the text, it was necessary to limit the coverage of the book, which was difficult, given the flexibility of the Excel environment. In the end, we tried to balance the range of topics with a suitable depth of coverage to enable readers either to use directly, the examples we include or to gain rapidly enough experience to apply the techniques we demonstrate to their own applications. The provision of sample case studies is a deliberate choice, as in our experience, analysis of detailed examples is the best way to pick up spreadsheet skills. We are of the opinion that the preparation of a spreadsheet solution to a data processing problem is a form of programming, and, as in traditional programming, the best way to learn is to examine programs written by experienced programmers. For this purpose, we are including sample spreadsheets and data files on disk for the user to play with. Have fun!

LIMITATIONS OF EXCEL

While Excel is a superb vehicle for accomplishing many data processing tasks for scientists, it cannot be expected to do everything easily or as well as other packages. As Excel users for many years, we are pleased to see what was primarily a business package develop to include many useful scientific and statistics tools (for example, the regression and trendline tools only appeared after many years). However, this is still an evolving process, and there remain limitations that should be appreciated. In our experience, the following points should be addressed in order to provide an improved environment for processing scientific data.

The 3-D display options are limited and not really of much use with real experimental data. Other packages, such as SIGMAPLOT, offer much better options in this area. However, the ability to nest SIGMAPLOT within an Excel spreadsheet is a recent development that should enable the user to take advantage of the better points of both packages.

The statistics options do not cover some of the more advanced tools recently developed (and which continue to be developed) for processing complex data sets (e.g., principal components analysis). Once again, a more specialized package, specifically directed to advanced statistics or chemometrics, is required.

The same situation arises in advanced engineering tools. For example, although Excel contains a Fourier Analysis option in the data analysis add-on, it is clumsy to implement and it is difficult to extract the processed information in the required format (the real and imaginary component of the transformed data are combined in single cells and must be separated in order to extract the real component). It is much easier to perform this type of operation in a more specialized engineering package such as *LabView*.

The user should be aware of the above design limitations in Excel. Despite these limitations, we enthusiastically recommend Excel as a means for learning how to process experimental data in a digital environment and for better understanding how modern software performs these tasks for us because, in contrast to the "black-box" approach, the user must program Excel to perform the appropriate tasks.

Dermot Diamond
Venita C. A. Hanratty

Dublin City University

ACKNOWLEDGMENTS

This book is the culmination of many years of teaching and research experience for us both. Our ability to produce it rises from knowledge passed to us through our teachers and experiences gained from our university colleagues and the students we have taught. It also rises from the encouragement and assistance provided by our families. Producing this text has required many long hours of preparation, which we could not have given without their active support. Thanks to Pat, Aisling, Mary-Margarett, Rachel, and Elliot and to Tara, Danny, and Helen.

We would like to thank the many colleagues who provided us with experimental data and useful advice during the preparation of this book: Dr. Francisco Saez de Viteri, Dr. Robert Forster, Dr. Brendan O'Connor, Dr. Ciaran Fegan, Dr. Conor Long, and Dr. Han Vos. We also acknowledge the influence of previous authors who have promoted the use of computer-based experimental data analysis and those who, over many years of endeavor, have developed the fundamental equations and algorithms used in our examples. Finally, we acknowledge the important contribution of the numerous undergraduate and postgraduate students who performed the bench work and whose constructive comments have encouraged us to produce this text. We particularly hope that this text will help to convince skeptics of the benefits of spreadsheets for processing and analyzing scientific data or will help those who are interested in using spreadsheets but are unsure of how to proceed.

USING THE DISKETTE

The diskette accompanying this book includes a program that will install the spreadsheet files that are discussed in the chapters. You must have a computer with Windows to use the included software. The files can be used with Microsoft Excel versions 5 and higher. The SPRDCHEM installation application uses the WINZIP program and will run under both the 16-bit Windows 3.1 system and the 32-bit WINDOWS 95 and WINDOWS NT systems. To install the spreadsheet files, run the SPRDCHEM.EXE program. The default install drive and directory is C:\SPRDCHEM, but you may edit this choice if you wish. Select the UNZIP option when you are ready and the files will be installed to your computer. To access the files, load your spreadsheet program and browse to the SPRDCHEM directory.

Some files have names that are longer than the 8.3 format required under Windows 3.1 and their names will be truncated if you install the files under that system. A table showing the names of the files in the 16- and 32-bit systems follows:

Chap	WINDOWS 95 Name	WINDOWS 3.1 Name
1	chap1	chap1.xls
2	regress	regress.xls
	statsdata	statsdat.xls

Chap	WINDOWS 95 Name	WINDOWS 3.1 Name
3	chap3	chap3.xls
4	chap4	chap4.xls
5	arrhenius	arrheniu.xls
	cu-nh3	cu-nh3.xls
	ethanoic	ethanoic.xls
	h3po4	h3po4.xls
	kinetics	kinetics.xls
6	feature extraction	feature.xls
	fia	fia.xls
	Mafilter	mafilter.xls
	IRspectrum	irspectr.xls
7	amylase	amylase.xls
	emghplc	emghplc.xls
	fluoexp1	fluoexp1.xls
	fluoexp2	fluoexp2.xls
	fluoexp3	fluoexp3.xls
	GAUSSIAN and noise	gaussi-1.xls
	Gaussian	gaussian.xls
	Gaussian2	gaussi-2.xls
	Ghplc	ghplc.xls
	malate	malate.xls
	PDAData	pdadata.xls
	selectivity	selectiv.xls
	Sigk@12r	sigk@12r.xls
	Sigk@6r	sigk@6r.xls
	Tnkhplc	tnkhplc.xls
8	Chap8	chap8.xls

CHAPTER 1

INTRODUCTION TO EXCEL

1.1. WHAT IS A SPREADSHEET?

A spreadsheet is a collection of data entries: text, numbers, or a combination of both (alphanumeric) arranged in rows and columns used to *display, manipulate, and analyze* data. Data organized in a table format is more efficient to manage. Excel prefers data to be vertically arranged, in columns rather than in rows. In this form, groups of cell entries can be operated on at the same time.
Excel is a spreadsheet software package that allows you to:

- Manipulate data through built-in or user-defined mathematical functions.
- Interpret data, for example, through graphical displays or statistical analysis.
- Create tables of numeric, text, and other formats.
- Graph tabulated input in various formats.
- Perform various curve fitting procedures, either built-in (linear or nonlinear regression) or user-defined through the powerful add-on *Solver*.
- Write user-defined macros or Visual Basic for Applications (VBA) routines to automate or enhance a spreadsheet for a particular purpose.

This chapter introduces routine spreadsheet operations to use in the case studies later in the book. More specific operations will be introduced in the case studies. The reader is referred to the *Microsoft Excel User's Guide* and On-Line Help for further information.

1.2. FUNDAMENTAL SPREADSHEET OPERATIONS

1.2.1. The Spreadsheet and the Excel Workspace

Launching Microsoft Excel brings you into the workspace of Excel and a new workbook. A workbook is a collection of related spreadsheets (worksheets or sheets). Each worksheet contains 16,385 rows (1 to 16,384) and 217 columns (A to IV) for a total of 3,555,545 cells. Rows have number headings and columns have letter headings. A cell is the intersection of a row and a column. Notice that the cell in the top left-hand corner is surrounded by a darker border than the other cells. This indicates that A1 is the active cell; it is selected.

The basic parts of a worksheet are labeled in Figure 1.1. These are workbook

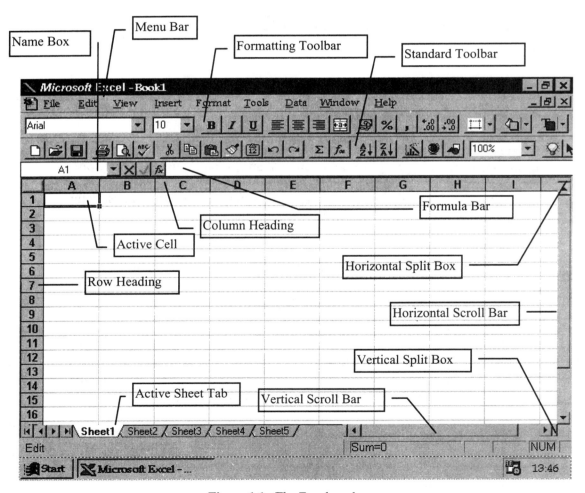

Figure 1.1. The Excel workspace.

name (default *Book1*), column heading, row heading, cell, vertical, horizontal, and tab scroll bars, worksheet and tab split box, active sheet.

At the bottom of the spreadsheet are sheet tabs representing the worksheets of the workbook. The default number of sheets each new workbook contains is 16. This number can be adjusted in Tools_Options_General tab shown in Figure 1.2. More sheets can be added (Insert_Worksheet) or deleted (Edit_Delet Sheet) as necessary. Click on these tabs one at a time to move from one sheet to another within the same workbook.

Tools_Options is where the workspace is customized to the user's preferences. Spend a few minutes exploring all the tabs. It is important to be familiar with their contents. Pressing the ESC key will exit any unwanted dialog box; that is, if you start to do something you don't want to do, ESC will save you further grief. Excel has a very efficient On-Line Help system. For further clarification of any command, just look it up in Help, double click the Help icon in the Standard toolbar, or click the Help button in the dialog box in question.

All the sheets will be empty as Excel has just been opened. The name of the workbook can be changed from *book1.xls* by saving it.

- Save now and enter the title **Chap1.xls.**

Notice that the file extension for Excel files is *.xls*. It is also good computing practice to give your files sensible names, so they can be easily found when needed. Excel 7 permits the use of long names, filenames of more than eight charac-

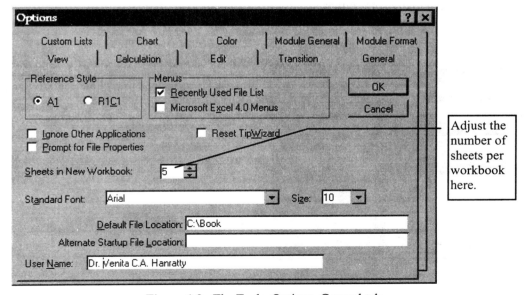

Figure 1.2. The Tools_Options_General tab.

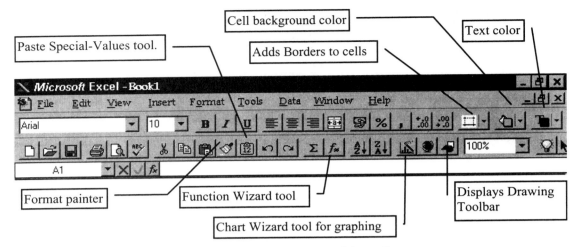

Figure 1.5. Command buttons of the toolbars.

the buttons of the Edit commands. To place the Paste Special Value button in the Standard toolbar just drag and drop it from the Customize dialog box to the desired position on the Standard toolbar.

Take note of the location of the Name box and Formula bar. The toolbar default setting is to have the Standard, Formatting, Formula, and Status bars displayed. The Status bar displays a description of the command pointed to or chosen.

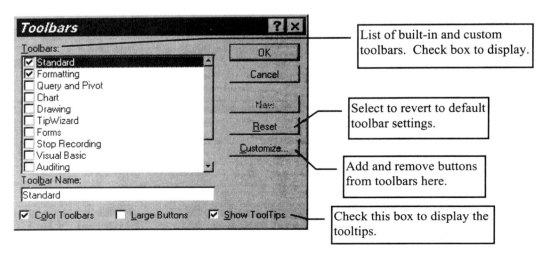

Figure 1.6. The Toolbar Settings dialog box.

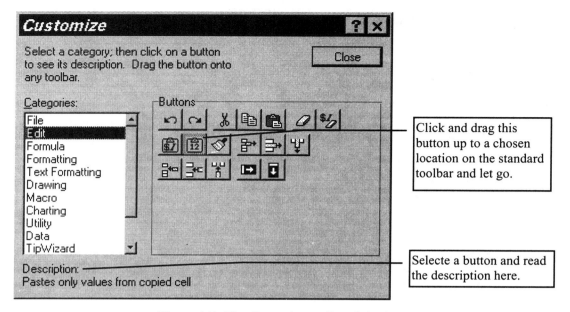

Figure 1.7. The Customize toolbar dialog box.

1.2.3. Moving around the Spreadsheet

Move around the sheet using either the mouse, arrow keys, a keyboard shortcut. As with any Windows package, you have the option of using the mouse or keyboard; however, for most instances, the mouse is more efficient. On a sheet you can move various ways. Look up *navigation* in Excel's On-Line Help for a complete list of options.

Case Study 1.1. Navigating

Try all of the following.

- Move the mouse over the desired cell and click.
- Move one window frame at a time with PAGEUP or PAGEDOWN.
- Scroll the horizontal and vertical scroll bars to move through large sheets. Press the HOME, CTRL + HOME, END + ↑ , HOME +→ keys and their combinations to see their response. (Alternatively, END and CTRL + END will bring you to the end of your sheet once it contains data.)
- Type the desired cell reference in the Name box. Select the drop-down arrow of the Name box and enter 1) **IV1,** 2) **A16384**.
- For a very large spreadsheet, two sections can be viewed at once; split the pane by double clicking the Split box and scrolling to the desired locations

in each pane. You need to click on the pane to activate it and then scroll. Another method is to use the View_Menu and select a lower magnification to view more of the spreadsheet at one time.

- Through Edit_Go To, enter specific characters or names for the search. Choose the Special button from this dialog box to view other possible modes of search.

Case Study 1.2. Navigating through a Large Spreadsheet

- For practice, split the window into two panes by either double clicking or dragging the horizontal split box.
- Scroll the top pane so cell A50 is visible.
- Then click the bottom pane and expose the bottommost cell of the sheet by using Edit_Go To and enter **A16384**.
- With the window split into panes, one pane can be frozen with respect to the other, from Window_Freeze Panes. Panes need to be unfrozen before the split can be removed.
- Double click the horizontal split box to return the sheet to a single pane.

As the active cell moves through the sheet, notice the cell address in the Name box also changes, corresponding to the new location. Spend a few minutes moving around the spreadsheet in the various ways until you are comfortable. Also observe the Name box to become familiar with how cells are denoted.

1.2.4. Cell References

Each cell has a unique address known as its reference. As noted, a cell is addressed by row and column headings. There are two reference styles by which a cell can be referenced: the row–column or R1C1 style and the A1 style. The cell reference *style* is set in Tools_Options_General tab (Figure 1.2): the default is A1. The exercises in this book are based on the A1 style. The current cell reference of the active cell is displayed in the Name box based on the selected style.

There are also cell reference *types*. Cell reference types are used in formulas, instructing Excel how to find the data to perform calculations. How the types differ in usage by mathematical formulas will be demonstrated in the next section. There are three reference types: *relative, absolute,* and *mixed.* Table 1 gives an example for each type for both styles. Placing the cursor in a cell containing a cell reference and repeatedly pressing F4 toggles through the various options of relative and absolute references. If a cell contains more than one cell reference, the cell reference where the cursor is placed will toggle. The use of either R1C1 or A1 reference type will usually depend on user preference.

Table 1.1. Cell Reference Types Used by Excel.

	R1C1	A1
Relative	R[2]C[3]	A3
Absolute	R3C1	$D$8
Mixed	R4C[1], R[−1]C2	E$5, $B2

Excel finds cells according to its reference type. If the reference is relative then Excel finds another cell by starting from the given cell. In R1C1 style, "positive" means move to the right or down in a row or a column, respectively. If it is absolute, a cell is located by the exact cell reference. A mixed reference cell address is a combination of both. The use of either relative or absolute reference types will usually depend on the task at hand, especially in reference to calculations.

Case Study 1.3. Reference Styles and Types

Here's a "Keep your thumb in your place and you won't get lost" exercise. Put your answers in relative reference type.

1. If the active cell is C3, what is the location of the cells, in A1 reference style, directed by each R1C1 reference type listed in Table 1? (*Ans.* F5, D6, D7, E2)
2. In order to get to D5, what is the notation in R1C1 reference style for each A1 reference type listed in Table 1? (*Ans.* R[2]C[3], R[−3]C, RC[−1], R[3]C[2])

1.2.5. Ranges and Arrays

Selecting a group of cells on a spreadsheet is done in the usual Windows software manner. To select A1 down to A12, click on the first cell and drag down to the last cell. Notice that while you do this operation the Name box changes accordingly to the selected cell. On completion, the Name box contains the cell reference of the first selected cell. This collection of cells is known as a range. A single cell can also be regarded as a range; in most cases, however, a range refers to more than one cell. The above selected range is denoted A1:A12 when used in the formula bar in functions. The colon can be interpreted as "to," hence, all cells from A1 to A12. In Excelspeak the colon is a reference operator producing a range of all cells between and including the two references.

There are two common techniques for selecting a range. The first is to high-

- Try this now by placing **10** in C3, then dragging the C3 fill handle down to C13. The value 10 is placed in C3:C13.
- Alternatively, select C3:C13 and enter in **10** (after first clearing the contents). Instead of pressing return, use the key sequence CTRL + RETURN. This places 10 in the selected range.
- Enter **10** in D3 and drag the fill handle with the CTRL key depressed. The numbers 10 to 19 have been entered.
- Enter **5** in E3, **10** in E4, and then select E3:E4. Drag the fill handle down a few cells. The entries are observing the already present trend.

AutoFill is convenient when entering weekdays, months, or any list into column headings.

- Enter **January 1995** into A2. Notice that what is entered may vary from what you typed. Excel displays the date in the format according to what is set in Format_Cells_Number(tab)_Date.
- Drag the fill handle to A13 and all the months of the year are entered.

The AutoFill entries come from either built-in lists or custom lists. These can be viewed in Tools_Options_Custom List tab. Custom lists can be created in this dialog box. View On-Line Help by searching for custom autofill lists for more information. A shortcut menu can be displayed to assist multiple cell entries by dragging the fill handle with rhmb depressed. With Excel 5 and 7, the rhmb is context sensitive, meaning when it is depressed, depending on the action being carried out, the appropriate shortcut menu will be displayed. This eliminates having to go up to the menu bar in many instances.

1.2.7. Editing Data

Editing data with the usual *cut, copy,* and *paste* operations are available in the Edit menu or from the shortcut edit menu by depressing the right mouse button on a selected cell or range. Cells need to be copied or cut before they can be pasted. A word of caution when editing: Microsoft Excel can only undo the last action; it is not sophisticated like its sister package, Microsoft Word, where multiple undos are possible. So go slowly when editing and save often! Unlike Microsoft Word, Microsoft Excel does not maintain cut or copied data in the Clipboard after another action is carried out. Maintaining the buffer is not a feature of the spreadsheet software but it is in the wordprocessor. Notice that there is a Paste Special option on the Edit menu. This is most helpful when pasting the results of a mathematical computation, as will be seen in the next section. There are other useful options in the Paste Special dialog box, especially the operations and transpose.

Moving data cells around the spreadsheet is a simple operation when rearranging data is required. Highlight the data in A3:A13 and then move the mouse to the edge of the selected area. The mouse cursor becomes an arrow pointing to the left. Click and drag to E5:E15. The CTRL key held down during dragging creates a copy of the selected data. Dragging with the right mouse button depressed brings up the shortcut menu on options when moving cells.

Inserting cells, rows, and columns is done through either the Insert or Shortcut Edit menu. Select B5 and bring up the Insert Cell dialog box through either menu. The new cell is created by shifting cells down or to the right. It is important when working with large spreadsheets that you ensure other areas of your sheet are not adversely affected by this edit. Excel inserts the new column to the left of the highlighted column. New rows are inserted above the highlighted row. To exchange rows or columns in a spreadsheet, highlight the selection, hold the SHIFT key down, and drag to the new location. Try this now by selecting the range C3:C13. Move the mouse to the edge of the selected area and hold down the SHIFT key. Drag this column between columns A and B and let go of the mouse button (while still holding down the SHIFT key) when the hatched vertical I-beam indicates a column to be inserted. The mouse button must be released before the SHIFT key for the move–insert action. If the SHIFT key is released first, a move–replace action is performed. Practice moving cells around the spreadsheet since this is an invaluable skill when results are needed in a hurry.

Further editing options are *delete* and *clear*. Deleting cells, rows, or columns will rearrange the data in the spreadsheet, so make sure that is the intention. The cells, rows, and columns are modified in the reverse manner in which they were inserted. Delete a few cells from your practice spreadsheet and see the effect. In many cases the desired action may be to clear the contents of cells. There are various options to Edit_Clear. When entered data sometimes does not appear as expected, clear the format of the cell and reenter if necessary (Edit_Clear_Formats).

Cells can be formatted in much the same way as in a wordprocessor software package such as Microsoft Word. The formatting options are found in the Format_Cells dialog box (Figure 1.9) and are arranged in tabs. Select each tab to familiarize yourself with the formatting options.

The Number tab allows setting the numerical data display from various formats along with other data formats such as time and date. An important aspect of formatting a spreadsheet containing scientific numbers is controlling the number of decimal places and precision. The latter can be done in Format_Cells dialog box or by the increase–decease decimal button in the formatting toolbar. The precision of calculation is taken as the total display of the calculated value to a maximum of 15 decimal places. The precision can be set for display in Tools_Op-

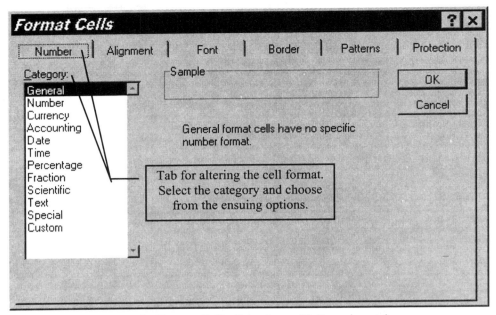

Figure 1.9. The Format_Cells dialog box with its various tabs.

tions_Calculation (tab). Search under *Precision* in the On-Line Help for more information. There is also a Round function for rounding off values that will be demonstrated in the next section.

To adjust the row height and column width, move the mouse pointer to between the desired headings; the mouse pointer becomes a "spacer" symbol. Click and drag the divider to the desired distance. This action can also be performed through Format_Column or Row. Double clicking the column or row spacer automatically adjusts to Autofit.

1.3. MATHEMATICAL OPERATIONS ON CELLS— ENTERING FORMULAS

1.3.1. Formulas

The fundamental operation of a spreadsheet is performing calculations on data. Excel performs mathematical operations through formulas and functions. Formulas are written by using the formula bar and beginning with an equal sign as shown in Figure 1.10.

For an equal sign to be displayed as the first character in a cell, it needs to be preceded with a single quote. The same is true for all mathematical operators.

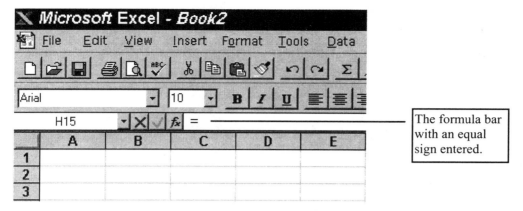

Figure 1.10. The Formula bar.

Functions can be either user-defined or built-in Excel functions. Since the formulas and functions are acting on cells of the spreadsheet, the variables used therein are the cell references of those cells. This is where the an understanding of the difference between relative and absolute references is important. Included in this section are various examples of using both.

Case Study 1.5. Temperature Conversion

- Enter the titles **Celsius** and **Fahrenheit** in cells A1 and B1, respectively.
- Adjust the column width so both titles are visible either through Format_Column_Autofit Selection or dragging the boundary between the columns with the mouse. A shortcut method is to double click this boundary.
- Format the titles to bold by highlighting the cells and pressing the *B* icon of the formatting toolbar.
- Enter the conversion range –10 to 150 from cell A2 down by Edit_Fill_Series.
- Enter **–10** into A2.
- Call up the Series dialog box and select series in **columns** and type **linear**.
- Enter step value **10**, stop value **150**, and press OK.
- In cell B2, enter the conversion formula from Celsius to Fahrenheit, starting with the equal sign (=), using the address A2 as the centigrade variable: **F = (C*9/5)+32**. When finished, press RETURN. **(= (A2*9/5)+32)**.
- Using the Split Pane tool, select the column B equivalent range to that of column A. The top pane needs to show cell B2. Scroll the bottom pane to reveal the end of column A data. This is shown in Figure 1.11 below.

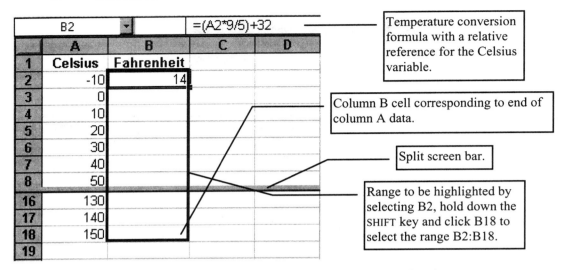

Figure 1.11. The split screen arrangement to aid selecting data.

- Select cell B2, hold down the SHIFT key and click B18. The cells B2:B18 will become highlighted.
- Use the Edit_Fill_Down command to copy your formula down through this highlighted range.

Click on some of the cells in the B2:B18 range and observe that the relative cell reference (A1 type) of the variable is incremented through this range (Figure 1.12). This is an example of relative referencing in formulas. If absolute referencing ($A$2) had been used as the variable in B2, this variable would have been constant throughout the calculation in the range in column B and the same answer displayed. Convince yourself that this is so by editing the B2 formula entry to the absolute reference and repeat the Fill_Down exercise.

1.3.2. Function Wizard

Excel's built-in functions are accessed through the Function Wizard tool, f_x, found on the Standard toolbar. Excel has many mathematical, statistical, and scientific functions. These have the general syntax:

$$= FunctionName(arguments)$$

For example, *SIN(number)* calculates the sine of an angle in radians and *SUM(range)* calculates the summation of the cells defined by the range. These

	A	B
1	**Celsius**	**Fahrenheit**
2	-10	=(A2*9/5)+32
3	0	=(A3*9/5)+32
4	10	=(A4*9/5)+32
5	20	=(A5*9/5)+32
6	30	=(A6*9/5)+32
7	40	=(A7*9/5)+32
8	50	=(A8*9/5)+32
9	60	=(A9*9/5)+32
10	70	=(A10*9/5)+32
11	80	=(A11*9/5)+32
12	90	=(A12*9/5)+32
13	100	=(A13*9/5)+32
14	110	=(A14*9/5)+32
15	120	=(A15*9/5)+32
16	130	=(A16*9/5)+32
17	140	=(A17*9/5)+32
18	150	=(A18*9/5)+32

Figure 1.12. Formula entries for temperature conversion.

functions are accessible through the Function Wizard, either from its tool button in the Formula bar, Standard toolbar, or the Insert_Function menu. Wizards are easy to follow, step-by-step procedures for the more commonly used tasks of Excel. To use a wizard, just follow the instructions in the dialog boxes, moving through them one step at a time. Figure 1.13 shows Step 1 of the Function Wizard. You can move back and forth through the steps of a wizard and also call up Help for further explanation in the use of the function.

Case Study 1.6. The Round Function

- Enter the constant π, 3.14159265, into A2 in a new spreadsheet. The value of π can alternatively be entered through the function =**PI( )**.
- Adjust the width of column A to reveal the complete entry by double clicking the column divider between columns A and B.
- Enter the Round function into B2, with its argument being the contents of A2, as described in the following steps.
- Click B2 and then put the cursor into the formula bar to display the Function Wizard icon (alternatively, double click B2 and it will be displayed).

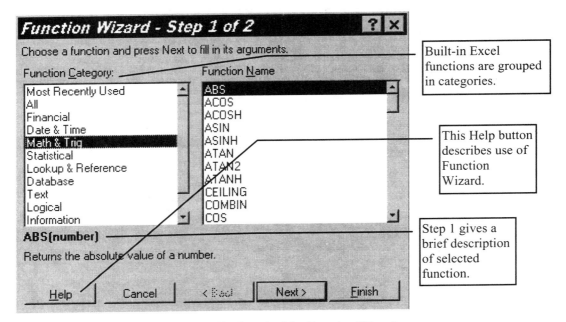

Figure 1.13. Step 1 of the Function Wizard.

- Depress the Function Wizard icon and in the first dialog box (Step 1) select **Math & Trig** from the Function Category list and **Round** from the Function Name list. Click Next.
- In Step 2 the first entry is for the cell address of the number to be rounded. This dialog box is shown in Figure 1.14.
- Notice that the cursor is already in this input box. You may need to move the dialog box to reveal cell A2 by dragging its title bar. This input box can be filled by selecting the cell A2 and the address will be entered. Press the TAB key once to move to the next input box named num_digits. In this example this is the number of digits after the decimal point. Enter **2**. Select the Help button to view other methods of rounding with the Round function.
- Press Finish and click the green arrow (or press RETURN) to calculate the result of the function. The result should be **3.14**.

Functions in Excel can be nested, which is placing one function within another. Generally this is more a requirement in logic decisions than mathematical calculations. However, they do have their purpose in scientific spreadsheets, for example, when formula calculations result in error messages. The classic instance is the divide by zero error (#DIV/0!). In many cases this error indicates a genuine

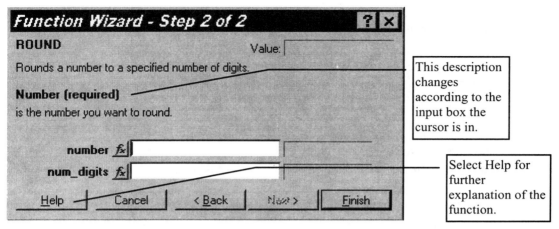

Figure 1.14. Step 2 of the Round Function Wizard.

problem with the calculation but in preformulated spreadsheets this error (or other error messages) may appear because all the data required by a formula are not in place. The case study below uses nested formulas to identify this situation and change the error message.

Case Study 1.7. Use of Nested Functions in a Predesigned Spreadsheet

- Select the next sheet tab (or insert a new sheet if necessary) and rename *is-error*.

- Enter titles **Mass C, Mass H, Mass O, Total Mass, %C, %H, %O** in A1:G1.

- In D2, enter the sum function to total the masses of C, H, and O [**=SUM(A2:C2)**]. This formula result displays a zero in D2. Zero values can be suppressed through Tools_Options_View (tab) in the Windows Options group; clear the check box next to **Zero Values**.

- Enter the formulas to calculate the percent composition of C, H, and O in E2:G2 (**%C=A2/$D$2*100**). These three formulas will give the **#DIV/O!** error message because the data has not been entered into the spreadsheet as of yet. When data for the masses of C, H, and O are entered, the formula will calculate accordingly.

- If this spreadsheet is to be used as a template, then it would be desirable to remove the divide-by-zero error message. This can be achieved by use of the IsError function nested in an If statement. The easiest method for demonstrating the construction of a nested function is through the Function Wizard. To begin, clear the formula entry from cell E2.

- Select E2 and choose the Function Wizard tool. From the Function Category select **Logical** and from Function Name select **If**, and choose Next.
- The logical test for the If statement in layman's terms is, "If the calculation is going to give a divide-by-zero error message, don't display it; display a blank cell instead." To achieve this, the IsError function is used as the logical test. Hence, select the Function Wizard tool to the left of the Logical Text input box.
- From this Function Wizard dialog box, select **Information** and **IsError**, then choose Next.
- In the IsError dialog box the value to be tested is the formula entered in the first instance: **A2/$D$2*100**. Choose OK.
- This brings back the If dialog box. Notice that double quotes have been placed around the argument of the IsError function. Remove the quotes so it reads: **IsError(A2/$D$2*100).** Tab to the second input box, **value_if_true**. Enter two double quotes, " ", the equivalent of a blank cell.
- Tab to the *value_if_ false* box and enter the formula as usual. Choose Finish. This nested formula now returns a blank cell when the data has not been entered into the spreadsheet. See Figure 1.15 for the spreadsheet thus far.
- Enter the nested functions for %H and %O in the same manner, making the appropriate changes for their respective masses.
- Enter data in cells A2:C2 for masses of C, H, O for ethanol, and their dependent cells will be automatically filled in. Figure 1.16 shows this final spreadsheet with the results of the formulas.

Check your formulas by comparing your percentages to those above. If cells A2:C2 are empty or have zero entered into them, the calculated cells E2:G2 remain blank. This type of formula allows preparation of a spreadsheet for data without the display of error messages when the data section is empty.

1.3.3. User-Defined Functions

User-defined functions are created through a Visual Basic for Applications (VBA) macro. An example of this is given in Chapter 3, Section 3.3. More exam-

	A	B	C	D	E
1	mass C	mass H	mass O	Total mass	%C
2	0	0	0	=SUM(A2:C2)	=IF(ISERROR(A2/$D$2*100),"",A2/$D$2*100)
3					

Figure 1.15. A nested **If** function to prevent display of an error message.

	A	B	C	D	E	F	G
1	mass C	mass H	mass O	Total mass	%C	%H	%O
2	24	6	16	46	52.173913	13.04348	34.78261

Figure 1.16. The Percentage Calculation spreadsheet.

ples of entering formulas and using functions will be given in the following section on graphing.

1.4. GRAPHING

Perhaps the common exercise performed by all scientists is graphing experimental data. Originally, Excel was not a good vehicle for performing this function, as the graphing options were clearly designed with the business market in mind, and many basic operations required by the scientific community were not offered. However, more recent versions have redressed this situation; Excel now can cope with most of the normal requirements and can offer some fairly advanced features such as the optimization add-on *Solver*, which, as we shall see in Chapter 7, can be used for quite advanced curve fitting tasks. While Excel does not challenge the features offered in specialized statistical and mathematical packages, it has very broad applicability and requires the user to enter the mathematical transformations to be performed. From the teaching point of view, it therefore provides a very powerful tool for teaching mathematics to undergraduate students, and for helping them explore graphically the importance of various parameters in an equation.

1.4.1. Chart Wizard

Graphs or charts in Excel are created through the Chart Wizard tool. This is an automated step-by-step guide to the entry of the required parameters for creating a graph. There are two types of graphs: embedded graphs, which lie on top of cells or other objects of the spreadsheet and a graph that is placed in a chart sheet. An embedded graph allows the user to view the graph and the data simultaneously. A chart sheet would be used whenever it is desired to have the graph on its own page. The first demonstration of graphing is generating an E (volts) vs. Time (sec) plot. The data is fabricated through the Edit_Fill_Series menu.

Case Study 1.8. Constructing a Graph

- Select the next sheet tab in the workbook. Double click the sheet tab and rename to *E vs Time*.
- Enter the titles **Time (sec)** and **E (volts)** in cells A1 and B1.

- Enter **9:00:00** in A2.
- Select the range A2:A21.
- Select Edit_Fill_Series and tick the **column** and **linear** choices. Enter a **step value of 00:00:30** to increment every thirty seconds.
- Enter **0.1** in B2.
- Select the range B2:B21.
- Select Edit_Fill-Series and tick the **column** and **linear** choices. Enter a **step value of 0.005.**
- The data is now ready to be graphed and the range is generally selected before using the Chart Wizard tool. Though this is not a necessary action to construct a graph, it does simplify the process. Selecting the range can be done either of two ways:
  - Select the range A2:B21 by click and drag starting from A2 down to B21.
  - Select the two column headings A and B.

This latter technique should only be used when no other data except that to be graphed are contained in the columns; otherwise editing the data series will be necessary. Select the data by either method.

- Click the Chart Wizard icon ▨ . The selection should have a moving dotted outline and the mouse cursor should now be a crosshair with a bar chart. This crosshair pointer is used to describe a box, by the click and drag technique, where the graph will be placed. Usually a graph is drawn on top of empty cells to the desired size, though not necessarily; this type of graph is an embedded graph and therefore an object that can be placed onto any cells or other objects.
- Click the mouse near the top left-hand corner of the area, hold down the left mouse button, and drag to the bottom right corner. Release the left mouse button.

The Chart Wizard steps (dialog boxes) appear on the screen. There are five steps in all. Selecting Next calls up the steps in sequence. There is also the option of moving back through the steps and calling on Help, as well as cancelling out of Chart Wizard and finishing early.

- In Step 1 the range to be graphed (Figure 1.17) is defined. Excel refers to graphed range as the data series. If the data range was not selected before choosing the Chart Wizard tool, it needs to be selected now by either of the techniques described or by typing in the range. Choose Next.
- Choose **XY (Scatter)** chart type by clicking in Step 2.

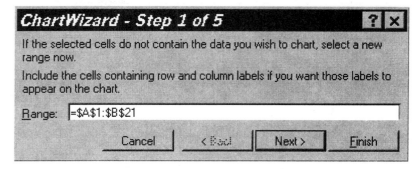

Figure 1.17. Step 1 of Chart Wizard showing range entry.

XY (Scatter) chart type gives the best results when working with scientific data. This type allows full manipulation of the x and y axes scales, and best calculation for the insertion of trendlines. (See Figure 1.18.) Choose Next.

- In Step 3 (Figure 1.19), for selecting a format, choose **format1**, without a line. Choose Next.

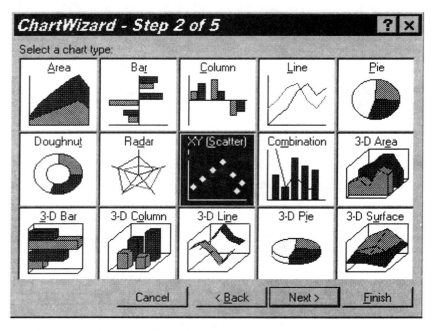

Figure 1.18. Step 2 of Chart Wizard for selection of chart type. XY (Scatter) is the desired type for most scientific data.

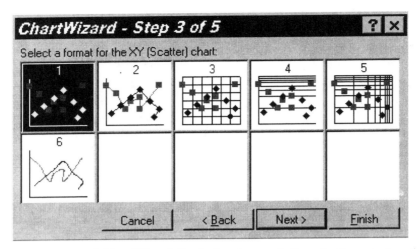

Figure 1.19. Step 3 of Chart Wizard displaying the graph formats. Formats 1 and 6 are useful. Format1 is best when curve fitting on data is to be performed. (Inserting a Trendline is demonstrated later in this section.)

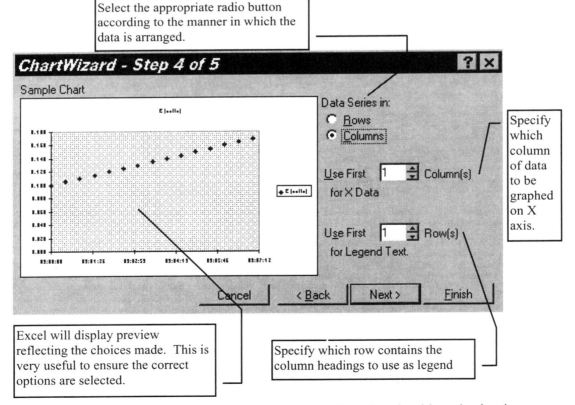

Figure 1.20. Step 4 of Chart Wizard. These options inform the wizard how the data is arranged on the spreadsheet. The wizard uses these to construct the graph.

- Step 4 determines the manner in which the data is graphed; this information is easily inferred from its layout in the spreadsheet (Figure 1.20). In this example, ensure **Data in columns** is selected. The further two selections in this dialog box are designation of the X data and the Legend. Excel defaults these two selections according to whether data is arranged in columns or rows. However, they can be changed to suit a particular graph. In this example, X data should be **column 1** and **rows** for legend. Choose Next.

- Step 5 allows the graph to be formatted in terms of Title and Axes Labels. Enter descriptions accordingly by tabbing through the input boxes. Pressing RETURN before filling in all the fields exits the Chart Wizard. These labels can be entered later by editing the chart (see later in this section). In option of a Legend, choose **Yes** or **No** as desired. Choose Finish or press the RETURN key.

The graph is now complete. See how your graph compares with Figure 1.22. If, when finishing chart wizard, the x or y axes scale looks too crowded

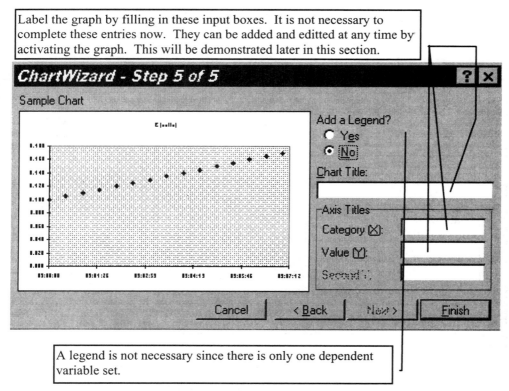

Label the graph by filling in these input boxes. It is not necessary to complete these entries now. They can be added and editted at any time by activating the graph. This will be demonstrated later in this section.

A legend is not necessary since there is only one dependent variable set.

Figure 1.21. Step 5 of the Chart Wizard allowing entries for labeling the graph.

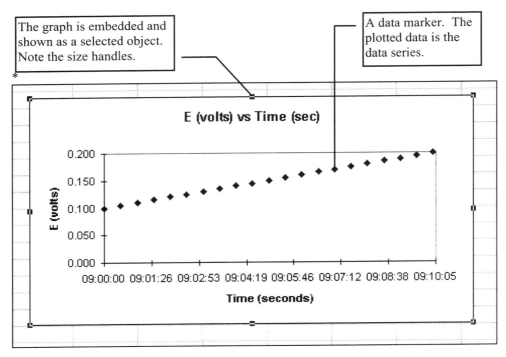

Figure 1.22. Graph of E vs. Time.

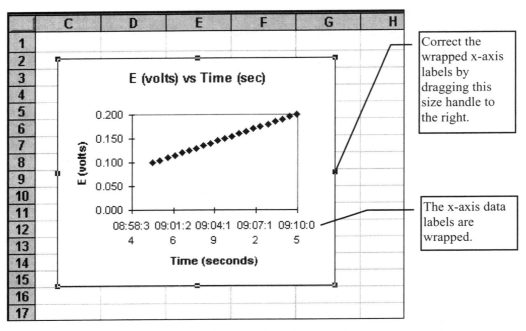

Figure 1.23. An embedded graph where the x-axis labels are wrapped.

26

or wrapped, it may just be that the initial graph area described was too small. Select the graph and drag one of the size handles to increase the size of the displayed graph object. This is shown in Figure 1.23. If the axes labels or chart titles were not entered in Step 5, they can be entered through the Insert menu when the graph is activated.

1.4.2 Graphing Mathematical Functions

This section reinforces techniques introduced in earlier sections through graphing typical mathematical formulas that are routine for Excel users from science backgrounds.

Case Study 1.9. Graphing a Simple Function

Let us examine a simple task: to graph the simple function

$$y = Ax^2 \tag{1.1}$$

over the range $x = 0$ to $x = 20$, in increments of 1. The constant A will demonstrate the use of an absolute reference in formulas.

- Open a new sheet and enter the titles **x** and **y** in cells A1 and B1, respectively.
- Enter the title **constant A** in C1, and enter **2** in C2.
- Enter the value **0** in cell A2 as the first value of x.
- Highlight cell A2; click on Edit_Fill_Series. Ensure **Columns** is selected and enter a Step Value of **1** and a Stop Value of **20**. Click on OK. The worksheet will appear with the x range automatically incremented by 1 over the required range. Note that this is a very useful way of entering series of data generally.
- Divide the screen by selecting the splitbox bar located just above the scroll-up arrow at the top of the scroll bar on the right-hand side of the screen (Figure 1.11).
- Drag the divider bar until the screen is divided into two parts, with the upper part being roughly twice that of the bottom. Scroll the bottom portion down until the final data in the series is visible (i.e., $x = 20$).
- Into cell B2 we will enter a function instead of a value. This is achieved by entering = as the first element in the cell. In this case, we want the cell B2 to contain the result of the calculation **Y = A*X^2** for the value $x=0$. Therefore enter = **\$C\$2*A2^2** into cell B2. The caret ^ is usually found

above the number six on the keyboard and represents raising the preceding number to the power of the following number. In this case, the preceding number is given by the address A2 (i.e., x = 0). Pressing the RETURN key executes the calculation and the result **0** appears in cell B2.

- Instead of typing in the above formula it is sometimes easier to click the cells to enter their cell references into the formula. In this case the process is as follows.
  - Type the equal sign.
  - Select cell C2.
  - Type the multiply operator,*.
  - Select cell A2.
  - Type in the caret, ˆ, and the number **2**. Do not press RETURN just yet. The C2 relative cell reference needs to be changed to absolute.
  - Move the cursor to the left to the C2 entry in the formula and press **F4**. This changes the cell reference to absolute. Continued pressing of **F4** toggles through all four variations of a cell reference, from relative to mixed to absolute and back to relative.
- The next task is to increment this formula over the range of x values. Select the range B2 to B22 by clicking on B2, holding down the SHIFT key, and clicking on B22 (the range should be highlighted). From the menu bar, select Edit_Fill_Down. This very useful command fills the range selected with the formula in B2 and automatically increments any addresses in the formula, meaning that the contents of each cell in the B column contains the square of the contents of each equivalent cell in the A column. Figure 1.24 shows cell B6 highlighted with the result **32** with the formula bar displaying its formula. Cells containing formulas normally display only the result of the calculation (as in this case), but when an individual cell is highlighted, the contents of the cell are displayed in the formula bar. The formula bar also provides a convenient way of editing the contents of a particular cell. Select various cells in the range B2 to B22 and confirm that the column A address has been automatically incremented. To graph the data we now perform the following actions.
- Select the range A1 to B22. Click on A1, hold down the SHIFT key and click on B22; the range should be highlighted.
- Click on the Chart Wizard tool. Move the mouse crosshairs and describe a graph by clicking and dragging from the top left-hand to bottom right-hand direction. The steps of the Chart Wizard appear on releasing the mouse button. Complete the steps as previously described above, ensuring that Steps 4 and 5 are filled in correctly. This graph is shown in Figure 1.25.

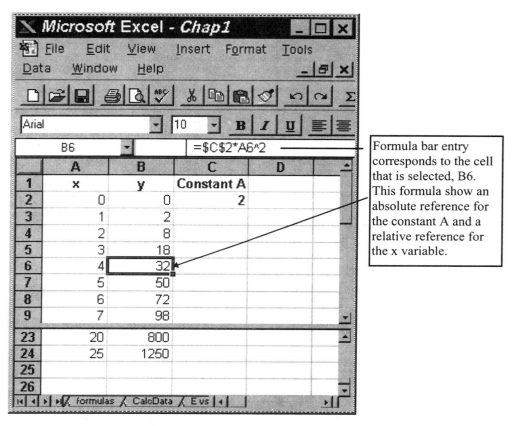

Figure 1.24. Spreadsheet for $y = Ax^2$.

1.4.3. Editing a Graph

A graph needs to be active to allow editing and performing further data analysis. When the graph appears, it will have a single black outline with boxes at the corners and midlengths of the sides. This represents a selected object. Figure 1.23 shows an embedded graph as a selected object. Double click inside the graph object to make it *active*. The graph outline changes to a hatched border indicating that the graph can now be edited. This is demonstrated in Figure 1.25.

Various chart attributes can be edited: labels, title, x and y axes, scales, data markers and so on, allowing further customizing and formatting of the graph. Text boxes can be added for further explanation of a graph. The chart format and type can be changed and the data can be altered. Once any object of a chart is selected, proceed to the Format menu and the first listing will pertain to that object. Alternatively, using the keypress CTRL + 1 once an object of the graph is selected will bring

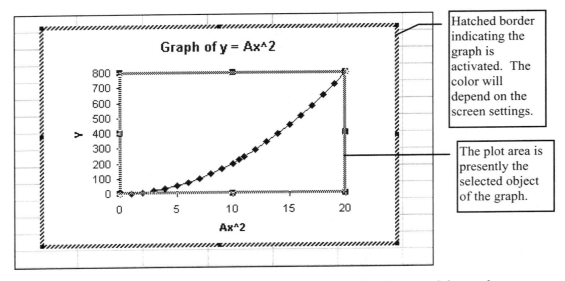

Hatched border indicating the graph is activated. The color will depend on the screen settings.

The plot area is presently the selected object of the graph.

Figure 1.25. An activated graph. Notice the hatched border around the graph.

up the appropriate Format Dialog box. The dialog box displayed presents all the options for editing that object. Other ways to bring up the Format dialog box are to click the right mouse button after selecting a chart object and the context sensitive shortcut menu will appear or to double click the object within the graph. In this section many of the ways in which a graph can be edited will be demonstrated.

Case Study 1.10.   Editing title and y-axis scale of the $y = Ax^2$ graph

- Activate the graph.
- Double click the graph title. The Format Chart Title dialog box comes up. Choose the Font tab and change the font to **bold**, size **10**. Choose OK.
- The title should still be selected. Click the mouse inside the title and change the text by highlighting it first and then typing the new title.
- Double click the y axis by moving the mouse pointer on the axis. If you have trouble doing the double click, select the axis once size handles should appear at each end of the axis. Now choose Format-Selected Object (or CTRL + 1). The axes are customized in this dialog box. Make sure you spend time examining the choices of each tab. Figure 1.26 shows the Scale tab.

Once a graph is created, the Chart Wizard can be called up again to edit the data series range and description. There are only two steps when the Chart Wizard is called up in this way, corresponding to Steps 1 and 4 in the usage of the wizard in the first instance. Data points can be added to the original range with ease. If

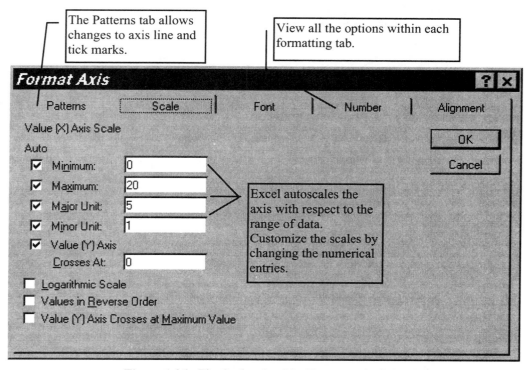

Figure 1.26. The Scale tab of the Format Axis dialog box.

the points are inserted within the existing range, the graph is automatically updated. If the new points are added to the range, then the Chart Wizard can be used to include them. The data series can be edited directly on the graph or the new points can be copied and pasted onto the graph.

Case Study 1.11. Modifying the Plot of y vs. Ax^2

Add a new x and y value within the A2:B22 range.

- Select row 12 by clicking the row heading and insert a new row.
- Enter a new x value of **9.5** in the blank row, in cell A12.
- Select cells B11:B12 and Fill-Down. This calculates the new y for the new x. The graph is automatically updated with the new point.
- Add a few new x and y values at the end of the range. Enter **22** and **25** into A24:A25. Select B24:B25 and Fill_Down.

The methods for adding new data points to the end (or beginning) of the range on the spreadsheet is demonstrated by three techniques.

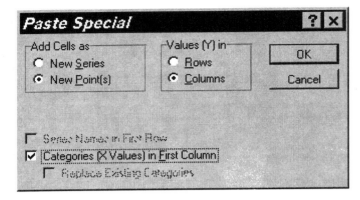

Figure 1.27. The Paste Special dialog box for pasting new cells to an existing graph.

1. Copy the new points, A24:B25 and copy (Edit_Copy). Select the graph and then choose Edit_Paste. A Paste Special dialog is displayed. In this example, Add Cells As **New Points**, Values (Y) in **Columns** and check Categories (X Values) in First Column. This is shown in Figure 1.27.
2. Select the graph and choose the Chart Wizard tool. Include these new points by editing the range of Step 1 to $A$1:$B$25 and choose Finish.
3. Activate the graph and select one of the data points. A series of data points is called the data series. The points become selected and the formula bar contains the data series entry:

   = SERIES(Sheet1!$B$1,Sheet1!$A$2:$A$23,Sheet1!$B$2:$B$23,1).

   Edit the A and B ranges from **23** to **25** and press RETURN. Notice that the graph is updated. The data series entry contains the sheet name where the data is located, in this example, *Sheet1*. The next two entries identify the x and y series comprising the data points and the last entry, 1, denotes there is only one data series in the graph. It is important to understand this notation when editing graphs with numerous data series.
   - Double clicking the data series on the graph calls up the Format Data Series dialog box. Move through the tabs and experiment the effect of changing the options therein.

1.4.4. Inserting a Trendline

Case Study 1.12. Performing a Linear Regression on the Data of the E vs Time Graph by the Action of Inserting a Trendline

- Select the *E vs Time* sheet and activate the graph.
- Click on the Y axis to select; square handles appear at either end.

- Now double click the axis and the Format Axis dialog box appears. There are five tabs for attributes that can be edited with respect to the axis.
- Select the **Scale** tab and increase the maximum by a small amount to have the plot wholly within the graph area. Choose OK.
- Click one of the data points to select the data series.
- Select the command Insert_Trendline.
- From the Type tab, select **Linear**. From the Options tab, select **display equation on chart** and **display R-squared value**. The Type tab is seen in Figure 1.28; it shows the built-in curve fitting types available in Excel.

The graph is complete. Selecting and double clicking various elements of the chart allows further customizing and formatting of the graph. Spend some time exploring these options. Many of these will be covered in the case studies.

The graphs obtained using the previous actions are usually not satisfactory in terms of appearance. However, you can edit the graph until you are satisfied. Furthermore, this completely customized graph can be stored as a *user-defined* format. This eliminates the need to repeat the same actions when constructing a graph of a similar nature. Excel also provides a means by which to apply different chart types on a previously constructed graph.

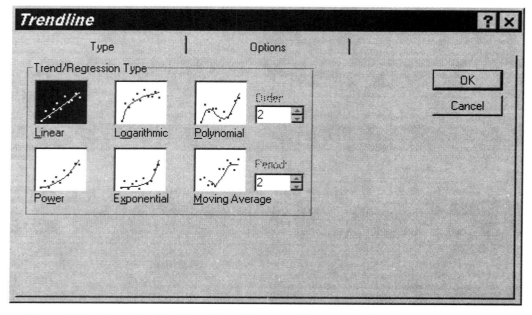

Figure 1.28. Type tab of the Trendline dialog box. These regression types represent those built-in in Excel. Excel can handle a variety of regression types. (The Options tab allows the choice of displaying the regression line.)

Case Study 1.13. How to Store a Format as a User-Defined Format and Apply When Needed

Use the temperature conversion spreadsheet for this demonstration. The aim is to create a general graph format for X Y scatter plots.

- Select the temperature conversion sheet and graph the data. Select **XY (Scatter)** chart type and **format1**.
- Double click on the graph to activate it. Graphs can only be edited when activated.
- Double click the *plot area* (inside the axes), and change the *area* color to **white**.
- Double click on the data series to edit the marker and line appearance. Change the line to **none** and the marker background color to **none**.
- The chart area can be resized within the window by clicking inside the axes and dragging the borders. Axes labels can similarly be edited and re-located anywhere in the graph window.
- Delete any title or axes labels. (Select them and press the DEL key.)
- Modify any graph attribute you choose as long as you remember the aim is to achieve a general graph format.

When the appearance of the chart is to your satisfaction, the format can be stored for future use in the following manner.

- Activate the graph (double click).
- From the Menu far, select Tool_Options_Chart (tab).
- Locate the **Default Chart Format** input box. The default is (*Built-in*). Click **Use the Current Chart** button and the *Add Custom AutoFormat* dialog box appears to enter a descriptive title for your custom graph format. The title can be up to 31 characters so do include reference to the format type, for example, **simple scatter.** Click on OK. This graph formatting has now been stored in Excel under the entered name. The Tools_Options_Chart (tab) also has variations for plotting empty cells. Spend some time experimenting with the options. User-defined formats are deleted via the Format_Autoformat dialog box, by selecting the User-Defined radio button and then the Customize button.

This custom format can be now be implemented for any graph via the Format_Autoformat menu. Remember, the graph must be activated to call up this menu item. This is also true for the Format_Chart Type and Tools_Options_Chart (tab). Select the **User-Defined** radio button and your custom format should appear in the list of available formats. Alternatively, one of the built-in chart format types can be applied to an existing graph. The following actions demonstrate how to implement a custom chart type.

- Select the *E vs Time* sheet tab and activate the graph.
- Choose ***Format_Autoformat*** and select the ***User-Defined*** radio button. Double click the custom format name created from the temperature conversion graph to apply it on the *E vs Time* graph.

A valuable user-defined graph format for scientific data is a simple X Y scatter graph since the majority of graphing is of this type. With this custom format defined, it can be set in Tools_Options_Chart (tab). A graph can quickly be created by finishing the Chart Wizard at Step 1 and the user-defined graph format set in Tools_Options_Chart (tab) will be applied.

This introductory chapter will close with one last example of the use of functions and graphing. This is another episode of reinforcing the skills you have picked up along the way and an opportunity to show you a few new techniques.

Case Study 1.14. Constructing a Sine Wave

Graph the sine function where x is in the range $0°$ to $1000°$.

$$y = \text{Sin}(x) \tag{1.2}$$

In Excel, the function **SIN(x)** returns the sine of the variable x, where x is in radians. Hence the first step is to convert degrees to radians. One way of approaching the task is the following.

- Enter the x range ($0°$ to $1000°$) in column A, incrementing every 10 degrees.
- Convert the degrees to radians in column B by entering the formula Radians = **degrees*π/180** in the first data cell of column B, substituting the first data cell in column A for *degrees*. Also, substitute **PI( )** for π.Fill down to convert the entire data set.
- Calculate the sine of the x by entering the formula = **sin(radians)** in the first data cell of column C, selecting the first data cell address for *radians* of column B in the formula. Fill this formula down in column C over the full data range.
- Graph the data as described above, but first the data needs to be selected. Your objective is to graph Column C vs. Column A. This can be achieved by selecting the entire Column B (selecting the column heading) and

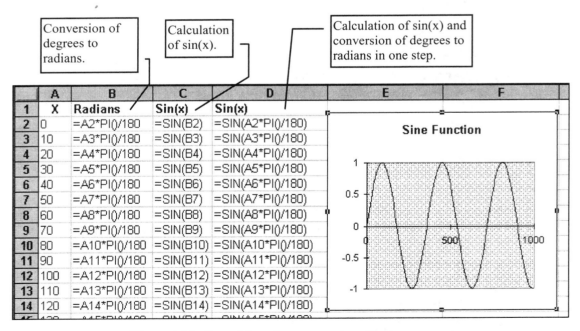

choosing **Format_Column_Hide**. Either split the window or drag to select the data of columns A and C. Choose **Edit_Go-To_Special** button and select **Visible Cells Only**. Now Chart Wizard is ready to be used. To unhide columns (or rows), select the column headings on either side of the hidden column (column A and C) and choose **Format_Column_Unhide**.

- An alternative method for selecting nonadjacent data for graphing is by use of the CTRL key. First select the data of column A in the usual manner: click on the first cell, hold down the SHIFT key, and click on last cell. The window will need to be split to view the last data of each column. To include the column C in the selection hold the CTRL key down and click on the first data cell of column C to highlight it. Release the CTRL key. Now press the SHIFT key and click on the last data cell in column C. The two data series should be highlighted. Graph the data using Chart Wizard.

The completed spreadsheet is shown in Figure 1.29.

Another way to design a spreadsheet to perform the calculation above it to eliminate the step of calculating radians from degrees as a separate column. Instead, incorporate this conversion in the sine function. The formula will then be = **SIN(A2*PI()/180)**. The spreadsheet now contains only two columns of data.

Figure 1.29. Spreadsheet for construction of a sine wave.

Taking an extra step in calculations by creating extra columns in spreadsheet development is a matter of preference. It has its advantages when troubleshooting calculations. Excel has a tool for examining the structure of calculations known as the Auditing tool. It is very useful in tracking errors in complex spreadsheets. Refer to the *Microsoft Excel User's Guide* for the discussion on the auditing tool in general. Here we will only look at its application. The auditing tool applies tracer errors for precedent and dependent cells in reference to formulas.

Case Study 1.15. Examining the Precedent and Dependent Cells of the Sine Wave Spreadsheet

- Select the sine function sheet.
- Select cell A3 and choose **Tools_Auditing_Dependents**. The cells that are directly dependent on A3 (or have A3 in their formula) are traced with a blue arrow. This is shown in Figure 1.30.

Selecting Tools_Auditing_Dependents again will trace indirect dependent cells on A3. Another way to trace dependents without arrows is to use the key press CTRL +]. Clear the arrows through **Tools_Auditing_Remove** all arrows. Reselect A3 and this time use the key press. The first time CTRL +] is used highlights the direct dependent cells and the second time highlights the indirect dependent cells.

- Now select D3. Choose **Tools_Auditing_Precedents** or use the key press CTRL + [. The cells contained in the formula of D3 are traced. This can be seen in Figure 1.31 below.
- Do the same with C3.
- Try this sequence of steps on the *IsError* sheet to reinforce the usage of the auditing, too.

	A	B	C	D
1	X	Radians	Sin(x)	Sin(x)
2	0	0	0	0
3	•10	0.17453	0.173648178	0.173648
4	20	0.34907	0.342020143	0.34202
5	30	0.5236	0.5	0.5

Figure 1.30. Dependent tracer arrows for the degrees entry of the sine wave spreadsheet. This spreadsheet shows only direct dependent tracer arrows.

	A	B	C	D
1	X	Radians	Sin(x)	Sin(x)
2	0	0	0	0
3	10	0.17453	0.173648178	0.173648
4	20	0.34907	0.342020143	0.34202

Figure 1.31. Precedent tracer arrows with C3 selected. This spreadsheet shows direct and indirect precedents.

1.5. PRINTING

There are a few items that will be mentioned here on printing in Excel 7. The Print Area menu item has returned to the File menu. This allows changing the print area quickly within a spreadsheet. Other options to note are found in the File_Page Setup_Sheet (tab). Here there is a Print group with helpful check boxes for customizing the look of the output. The Gridlines and Row and Column Headings are useful options. To print an embedded graph separate from the spreadsheet, activate it and then choose File_Print. For further information on printing, look up "Printing" in the On-Line Help.

1.6. SUMMARY

This first chapter has demonstrated many aspects of Excel. It was not the intention to cover all elements of Excel but to highlight through science-related examples those most commonly needed by users in the scientific community. The reader is advised to consult the *Microsoft Excel User's Guide* and On-Line Help for further information on topics discussed here.

CHAPTER 2

STATISTICAL FUNCTIONS
AND REGRESSION ANALYSIS

In this chapter, we shall introduce some of the statistical functions available in Excel and give some examples of their use. However, a detailed discussion on statistics and its application in processing scientific data is beyond the scope of this book. Readers are referred to more specialized texts such as *Statistics for Analytical Chemistry*, [1] for more information and data sets that can be used in order to become familiar with the application of these functions. The reader should also note that detailed help is available from within Excel via the Function menu. Extensive explanations of the functions, including examples and mathematical equations, can be easily accessed by the user.

2.1. STATISTICAL FUNCTIONS OF THE FUNCTION WIZARD

Statistical functions are selected from two menus within Excel. The first is through the Function Wizard and the second is from **Tools_Data Analysis.** Activate the formula bar of any cell and click the Function Wizard button ![fx]. From the Function Category list select Statistical (see Figure 2.1). The Function Name list box displays the available statistical functions of the Function Wizard. Scroll through the list and view the Excel keyword names. An explanation of the keyword is found in the comment area at the bottom of the Function Wizard dialog box. The arguments required for use of the function are displayed in the syntax statement also found at the bottom of the dialog box. For further clarification on the definition and usage of a function, select the Help button after selecting the function name in question (Figure 2.1).

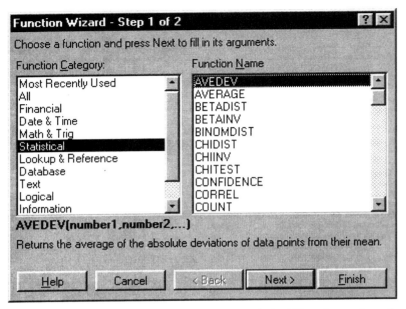

Figure 2.1. Excel Function window with Statistical Functions selected.

Example 2.1. Calculation of the Mean and Standard Deviation of a Set of Replicate Measurements.

Table 2.1 gives values for blood sodium in mM obtained from 10 replicate measurements of a single sample. Calculate the mean and standard deviation of the results.

Method

- Enter the sodium concentrations into a spreadsheet. Add comments to identify the data as appropriate;
- To calculate the mean, select a free cell and enter the formula *=average(range)* where "*range*" is the array of cells containing the data (e.g., A2:A11)
- To calculate the standard deviation, select a free cell and enter the formula *=stdev(range)* using the same range as above. Alternatively, one could use

Table 2.1. Replicate Measurements of Sodium Concentration in Blood

$n =$	1	2	3	4	5	6	7	8	9	10
Na/mM	140.29	139.09	139.50	139.84	139.24	140.06	140.25	139.25	140.69	139.01

the Function Wizard to access the functions after selecting the destination cell for the function. (Answers: mean = 139.72 mM; standard deviation = 0.58 mM)

Note: It is possible to devise questions of this type very quickly in Excel using the rand_between function which is available from the Function Wizard. Significant figures can be controlled by making the values f orders of magnitude larger than required and dividing the values by f orders of magnitude or through the **format_cell_number** command. Other popular statistical functions include *skew, count, median, var* (variance), *confidence, max,* and *min.*

For example, the confidence interval for the above data can be quickly calculated inserting the function =*confidence(alpha, SD, n)* into an unoccupied cell, where alpha equals the confidence interval (e.g., 0.05 = 95% confidence interval), SD equals the population standard deviation, and *n* equals number of data values. Hence in this case, one would enter =**confidence(0.05, 0.58, 10)**. The answer obtained is 0.38 mM, meaning that the true value of the blood sodium concentration is 139.72 ± 0.38 mM with a confidence of 95%.

The above example and data can be found in the sheet *blood sodium* in the workbook *statsdata.xls.*

Example 2.2. Comparison of Means Using the t-Test ([1], p. 57) and Standard Deviations Using the F-Test

Table 2.2 shows two sets of data for the analysis of thiol in the lysate of normal and rheumatoid patients (The data are available in the sheet *thiol data* in the workbook *statsdata.xls.*)

Table 2.2. Thiol in the Lysate of Normal and Rheumatoid Patients

	Normal	Rheumatoid
	1.84	2.81
	1.92	4.06
	1.94	3.62
	1.92	3.27
	1.85	3.27
	1.91	3.76
	2.07	
n=	7	6
mean	1.921	3.465
SD	0.076	0.440
t-test	—	−8.5

Method

Enter the data into a spreadsheet, add comments, and format as desired. The t-test formula can be quickly set up from the values of the means $(\bar{x})$, standard deviations (s), and counts (n) using the formula

$$t = \frac{(\bar{x}_1 - \bar{x}_2)}{\sqrt{(s_1^2/n_1 + s_2^2/n_2)}} \,. \qquad (2)$$

- Use the formulas *count(range), stdvev(range),* and *average(range)* to calculate the number of values, the standard deviations, and the means of both sets of data.
- Use these values to construct the t-test formula (equation 2.1).

From the magnitude of the t-value (8.5), we can conclude that, as it is greater than the critical value of $|t|$ (P = 0.05) = 2.57 from statistical tables [two-tailed test, degrees of freedom (df) = 5], the null hypothesis (that the two means are the same) is rejected and the means are statistically different.

The F-test is easily calculated from the ratio of the variances ($F = s_1^2 / s_2^2$) with the data series (1) and (2) arranged so that F is greater than 1 [in this case, (1) is rheumatoid and (2) is normal]. The calculation returns the value F = 33.955, which is much larger than the critical value obtained from tables [$F_{5,6}$ (P = 0.05) = 4.387] Note that the subscripts 5, 6 refer to the degrees of freedom (df). The conclusion therefore is that the precision of the two methods is significantly different.

2.2. Statistical Functions from Data Analysis ToolPak

The Data Analysis Tools are found under the Tools menu. If Data Analysis is not present when the Tools command is selected, this means the Analysis ToolPak was not loaded during Excel installation. If so, run the Excel Setup program again to install it.

The exercises in Examples 1 and 2 describe the traditional manner of analyzing data in a spreadsheet. However, the Analysis ToolPak provides a very fast method of analyzing and reporting statistical data. Let us analyze the data in Table 2.2 again, using this approach to demonstrate the power of the Analysis ToolPak and report generator.

Example 2.3. Using the Data Analysis ToolPak for Generation of t-Test and F-Test Reports

Method

- Open the sheet *thiol data* in the workbook *statsdata*.

- Select **Tools** in the menu bar and select **Data Analysis.**
- Select **F-test Two-Sample for Variances** and click OK. The dialog box shown in Figure 2.2 will appear.
- Enter Variable 1 and Variable 2 ranges (remember that the final value of F should be greater than 1—if you find it is less than 1, just repeat the calculation and reverse the variable identities; in this exercise, variable 1 is *rheumatoid* and variable 2 is *normal*).
- Alpha defaults to 0.05—leave this set.
- Under the Output options, select **New Worksheet Ply** so that the report will be generated on a new worksheet and click OK.

The formatted report shown in Table 2.3 is generated, which contains all the required information including significance values (no more tables to look up!). Obviously, this information can be very quickly assimilated into reports and papers for publication or circulation.

The t-test can be similarly implemented via the Analysis Toolpak using the same method as for the F-test (scroll down the options under Tools+Data_Analysis and choose t-Test: Two-Sample Assuming Unequal Variances). A dialog box similar to the F-test will appear and the same entries made. A report similar to that shown in Table 2.4 will be generated. Once again, all the information required to interpret the data is given and can be easily incorporated into reports and documents.

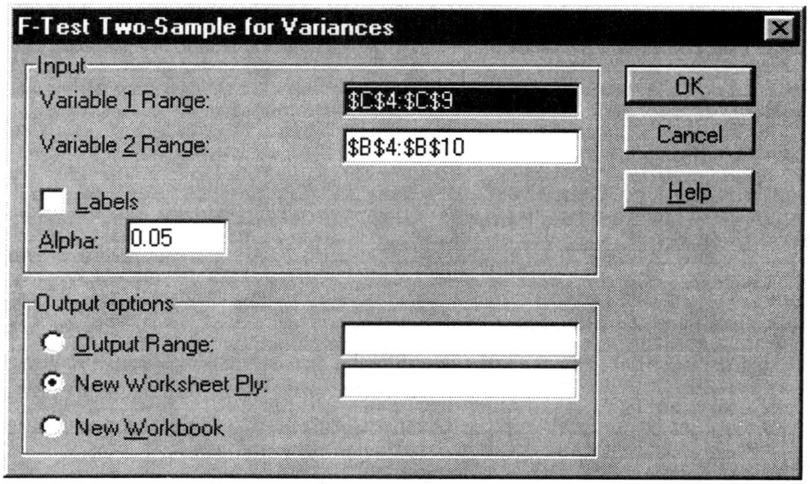

Figure 2.2. F-Test dialog box available under the Analysis ToolPak.

Table 2.3. Report Generated by F-Test Using the Analysis Toolpak

F-Test Two-Sample for Variances

	Variable 1	Variable 2
Mean	3.465	1.921429
Variance	0.19403	0.005714
Observations	6	7
df	5	6
F	33.95525	
$P(F \Leftarrow f)$ one-tail	0.000251	
F Critical one-tail	4.387374	

Table 2.4. t-Test Report Generated Using the Analysis Toolpak

t-Test: Two-Sample Assuming Unequal Variances

	Variable 1	Variable 2
Mean	1.921429	3.465
Variance	0.005714	0.19403
Observations	7	6
Hypothesized Mean Difference	0	
df	5	
t Stat	−8.47724	
$P(T \Leftarrow t)$ one-tail	0.000188	
t Critical one-tail	2.015049	
$P(T \Leftarrow t)$ two-tail	0.000375	
t Critical two-tail	2.570578	

Example 2.4. Use of Data Analysis ToolPak for Anova

The following example is taken from [1], p. 66. The stability of a fluorescence reagent was investigated by measuring the fluorescence (arbitrary units) from solutions stored under different conditions for 1 hr compared to the fluorescence of a freshly prepared solution. Each measurement was made in triplicate. The task is to determine whether the variation in fluorescence arising from storage time and/or storage conditions is significantly greater than the variation in the analytical method. The results are summarized in Table 2.5 and are available in the sheet *anova* in the workbook *statsdata.xls*.

Table 2.5. Fluorescence Signal Data

Conditions	Test 1	Test 2	Test 3
Freshly prepared	102	100	101
Stored 1 hr in dark	101	101	104
Stored 1 hr in low light	97	95	99
Stored 1 hr in bright light	90	92	94

Method

- Enter the data into a workbook, or open up the sheet *anova* in the workbook *statsdata.xls*.
- Select Tools+Data Analysis.
- Select Anova: Single Factor from the list of options—the dialog box shown in Figure 2.3 will appear.
- Enter the input range covering the entire data set; in Figure 2.3 this is **$B$2:$D$5**, which means that the first data value is in cell B2 and the last value is in cell D5.
- Enter whether the data are grouped in columns or rows (rows in this case).
- The Labels in First Column option should be checked if the labels (head-

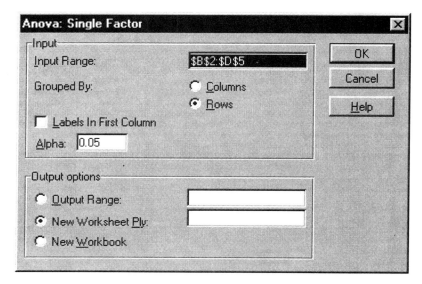

Figure 2.3. Dialog box for Anova: Single Factor available under the Data Analysis ToolPak.

ings) were included in the range. This is useful for keeping track of the data in the report.

- Alpha can be set (default is 0.05).
- Select the output range (new worksheet is useful).
- Click on the OK button.

The Anova report is automatically generated as shown in Table 2.6. The F-test result at 20.67 is greater than the critical value (4.07), meaning that the variance between the different samples is significantly greater than the variance within the measurement technique (random variance), and therefore, the material is not stable. Further tests are necessary to identify the source of the instability, which usually involve splitting the data set down into smaller units and comparing their variances. As there is little difference between the material stored for 1 hr in darkness and the fresh material, it would appear that the time factor is less significant than the brightness of light under which the material is stored. Further investigation is now required to reveal whether this factor is statistically significant.

In summary, these few examples demonstrate the power of Excel as a teaching tool. Students can be requested to input statistical formulas manually (in itself, this reinforces to students how statistical parameters are related to each other and ensures that the formulas must be properly entered into the sheet). On the other hand, the Analysis Toolpak provides a very powerful tool for the busy scientist who may have large amounts of data to process and reports to generate. Other statistical tools available under the toolpak are listed in Table 2.7.

Table 2.6. Anova Report Generated Using the Single Factor Anova Option under the Data Analysis Toolpak

		Summary		
Groups	Count	Sum	Average	Variance
Row 1	3	303	101	1
Row 2	3	306	102	3
Row 3	3	291	97	4
Row 4	3	276	92	4

Anova						
Source of Variation	SS	df	MS	F	P-Value	F Crit
Between Groups	186	3	62	20.66667	0.000400152	4.06618
Within Groups	24	8	3			
Total	210	11				

Table 2.7. Statistical Tools Available under the Analysis Toolpak

ANOVA: Single Factor	Descriptive Statistics	Correlation
ANOVA: Two-Factor with Replication	Covariance	Histogram
ANOVA: Two-Factor without Replication	Regression	Sampling
F-Test: Two-Sample for Variance	Rank and Percentile	Fourier Analysis
t-Test: Paired Two Sample for Means	Moving Average	
t-Test: Two-Sample Assuming Equal Variances	Random Number Generation	
t-Test: Two-Sample Assuming Unequal Variances	Exponential Smoothing	
z-Test: Two-Sample for Means		

2.3. Linear and Nonlinear Regression

Trends in data generation, due in part to the computerization of instrumentation, have led to larger and more complex data sets. Traditional approaches to experimental data processing are largely based on linearization and/or graphical methods. However, this can lead to problems where the model describing the data is inherently nonlinear (i.e., cannot be linearized), or where the linearization process introduces data distortion (e.g., in standard deviations of data that are logarithmically related to the analyte concentration, as with ion-selective electrodes). Hence, with the ready availability of PCs in laboratories, there is increasing interest in applying nonlinear curve fitting techniques to experimental data.

Computers provide an ideal tool to overcome such problems. Although statistical programs have been available for large mainframe computers for many years, these were often cumbersome to use and the computers were not readily available. Many of these statistical packages have now been converted for PCs, but their application to solving analytical data processing problems is still relatively rare because, although scientifically correct and effective, these packages tend to be engineer-orientated, not especially user-friendly, and expensive. With the increasing availability of these regression tools, one would assume that the standard of data analysis would dramatically improve compared to the past, when regression equations had to be laboriously built up from a series of repetitive calculations. However, the opposite is often the case, since students tend to use curve fitting tools uncritically, even when a cursory visual examination shows the fit to be unacceptable. Common examples include fitting a linear regression equation to data that are clearly nonlinear in character or fitting a polynomial that passes through all the points, but does not follow in any manner the overall trend in the points.

Excel provides some built-in tools for fitting models to data sets. By far the most common routine method for experimental data analysis is linear regression, from which the best-fit model is obtained by minimizing the least-squares error between the y-test data and an array of predicted y data calculated according to a linear equation with common x-values. There are several ways in which linear regression can be accessed in Excel, three of which will be examined below.

2.4. The LINEST Function

The LINEST function produces an array of output, that is, generating a matrix or table of output. However, this output matrix is only created through entry of the function as an array formula. The size of the output matrix depends on how large a matrix is highlighted before entering the LINEST function in conjunction with its declared arguments. This will be demonstrated in the subsequent examples.

LINEST allows for detailed multiple linear regression and the template for the returned output array of this function is shown in Table 2.8. The subscript n represents the number of multiple independent x-data sets.

The abbreviations should be familiar: m is slope, se is standard error values on the slopes, r^2 is the coefficient of determination (variance), se_y is the standard error in estimating y, F statistic, df is degrees of freedom, ss reg is the regression sum of squares and ss $resid$ is the residual sum of squares.

Case Study. Using the LINEST Function

- Select a blank sheet tab and rename to linest.
- Enter the data of Table 2.9 in A1:B14.
- Enter **LINEST** in cell D1.
- Activate cell E1, select **LINEST** from the Function Wizard and enter the arrays for the known_ys and known_xs accordingly. Leave the const and stats input boxes blank or cleared. Choose Finish. The LINEST Function Wizard dialog box is shown in Figure 2.4.
- The cursor will be in the formula bar. Click the green checkmark or press RETURN on finishing the dialog box. The value in E1 is the slope of the regression line.

Table 2.8. LINEST Output Array Format

M_n	m_{n-1}	. . .	m_2	m_1	b
se_n	se_{n-1}	. . .	se_2	se_1	se_b
r^2	se_y				
F	df				
ss reg	ss $resid$				

Table 2.9. Test Data Set for Linear Regression

x-index	Test Set
	0.126
2	0.202
3	0.304
4	0.355
5	0.505
6	0.530
7	0.600
8	0.648
9	0.757
10	0.790
11	0.858
12	0.899
13	0.935

To Display the Slope and Intercept

- Enter **LINEST** {m,b} in cell D3.
- Highlight cells E3:F3 (a 1 × 2 matrix), select **LINEST** from the Function Wizard and enter the input as before.
- Do not press RETURN after leaving the dialog box. Instead, the formula needs to be entered as an array formula. There will be a value present in cell E3

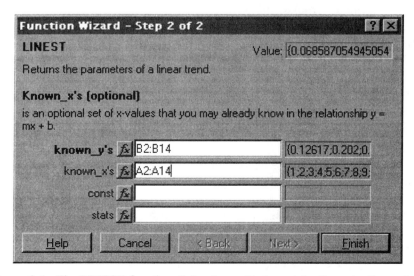

Figure 2.4. The LINEST function dialog box with the entries from the linest sheet.

and the E3:F3 range will be highlighted. Click the mouse in the formula bar and place at the end of the entry if it is not already there. Now press CTRL + SHIFT + ENTER. The values of the slope and intercept have been entered into cells E3 and F3, respectively.

The Complete set of Regression Parameters

- Enter **LINEST o/p** in cell D5. Adjust the width of column D by double clicking the column divider between columns D and E.
- Highlight a 5 × 2 matrix starting with E5, that is, the range E5:F9. Only two columns are necessary as there is only one set of independent x values.
- Select **LINEST** from the Function Wizard and enter the known_ys and known_xs as before.
- Enter **TRUE** in the const input box instructing the intercept to be calculated. Entering FALSE would have required the regression line to pass through zero by setting the intercept to zero.
- Enter **TRUE** in the stats input box. This logical value specifies the additional regression statistics to be returned.
- After leaving the dialog box press CTRL + SHIFT + ENTER to enter the formula as an array formula. Do this in the manner described above. The full complement of regression parameters are displayed in the matrix E5:F9. The parameters of this array are as described earlier in this section.

This resulting output matrix is a unit; that is, no part of the array can be edited. It can only be edited or moved in its entirety.

If a reduced set of statistical parameters of the linear regression is desired then the matrix containing those statistics only need be highlighted before entering the LINEST function. For example, in a particular linear regression, the parameters required are slope, intercept, and the variance. The minimum matrix that displays these parameters (along with other parameters) is a 3 × 2 matrix.

- In D11 enter **LINEST red. o/p**.
- Highlight E11:F13.
- Enter the arguments of the LINEST function as above and press CTRL + SHIFT + ENTER after leaving the dialog box.

It can be seen that the LINEST function is versatile in its application. Moreover, Excel offers the SLOPE and INTERCEPT functions if limited data on regression is desired. It is also worthwhile to examine the TREND and FORECAST functions; these are similar to LINEST. Select them from the Function Wizard and choose Help for further information on how they operate.

2.5. Linear Regression: Access Via Data Analysis_Tools

The most detailed linear regression analysis is obtained via the Tools command in the menu bar. This is illustrated in the workbook *regress.xls* with the test data set given in Table 2.9. As an exercise, these can be entered into a workbook or copied from *regress.xls* to a new workbook.

**Example 2.5. Linear Regression of Data in Table 2.9 Via Data Analysis
ToolPak**

Method

- Under the Tools command, select the **Data Analysis** option and then select **Regression.** This opens a dialog box as shown in Figure 2.5.
- Enter the cell ranges for the x-index array and the test set array in the appropriate boxes (or click and drag the mouse over the data set on the workbook with the menu field active).
- Enter the top left cell of the region of the workbook where you want the re-

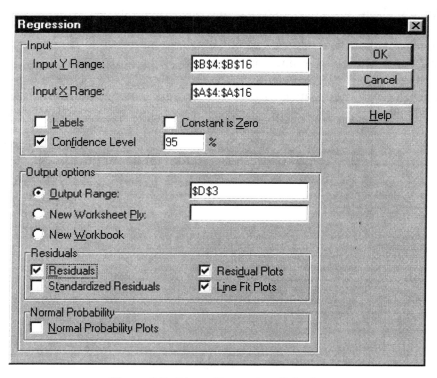

Figure 2.5. Linear Regression dialog box.

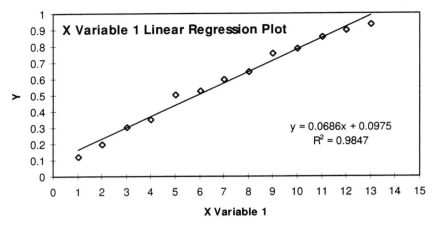

Figure 2.6. Typical linear regression plot.

gression analysis report to be copied (this should not coincide with any part of your data set—you can also direct the output to a new worksheet).

• Finally, check the options to display the Regression Line and Residuals Plot.

The regression fit and the residual plot obtained are shown in Figure 2.6 and Figure 2.7, respectively. The linear fit at first glance might appear to be reasonable, and the correlation coefficient, at $R = 0.9847$ for 13 observations is in itself not bad. However, examination of the residuals (difference between equivalent predicted data and test data points) shows some structure. For a good fit between

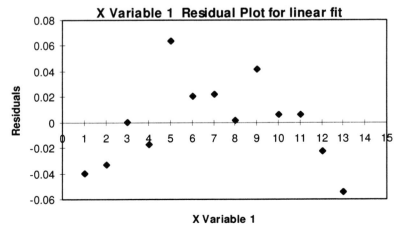

Figure 2.7. Residuals plot for Figure 2.6.

the two sets of data, the residuals should be evenly scattered about zero along the entire range of the arrays. Looking at the residual plot (Figure 2.7), the residuals initially are negative, go positive in the middle of the range, and then become negative once more at the upper end of the range, suggesting an underlying curvature. In any fitting exercise, examination of the residuals is important for identifying any underlying structure in the test data not described by the model.

This method of implementing linear regression also generates a detailed statistical report, which includes the slope and intercept coefficients for the best-fit line, the standard error in these coefficients, correlation coefficient, and a listing of the residuals.

2.6. Linear Regression Using Insert Trendline

A simpler approach than that described above is to use the *Insert Trendline* option. The best way to do this is as follows.

- Plot the experimental data using the Chart Wizard as described previously. Select scattergraph without joining points as the display format (i.e., just plot the data points).
- Activate the graph by double clicking anywhere inside it.
- Select the data by double clicking on any of the points.

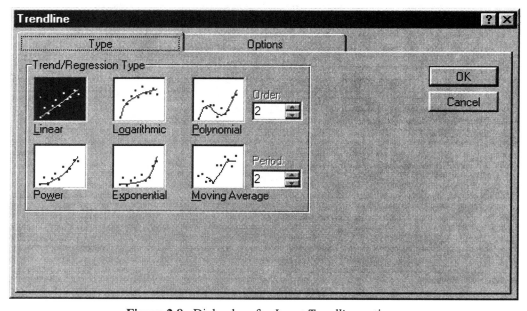

Figure 2.8. Dialog box for *Insert Trendline* options.

- Go to the Insert command and select ***Trendline***. This will open the dialog box shown in Figure 2.8. Note that there are several options available for nonlinear fits, but we shall ignore these for the present.
- Select **Linear** and then click on the options tab to open the dialog box shown in Figure 2.9. The options are as follows.
- *Enter Trendline Name*—this allows the user to override the automatic name applied to the regression line;
- *Forecast*—this allows the trendline to extended on either extreme (forward or backward) beyond the range of the x array. Note that this option is used in Chapters 6 and 7 to locate intercepts graphically.
- *Set Intercept to zero* allows the user to override the default linear model that includes an intercept;
- The Display Equation and Display *R*-squared *value* options, perhaps the most useful, automatically places these on the chart.

This approach enables a graph such as shown in Figure 2.5 to be generated very easily. Note that if similar data sets are to be analyzed, then the process can be speeded up using templates for standard graph formats, macros, or copying the spreadsheet and then pasting each new data set over the preexisting set. This last option is very simple to implement but applies only to data sets gathered against

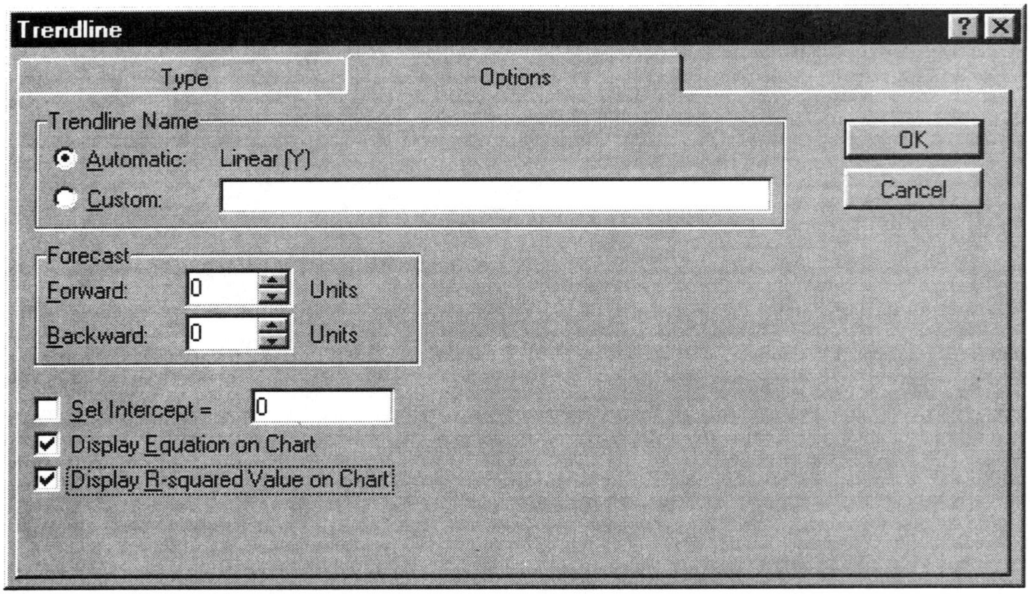

Figure 2.9. Linear regression options.

an identical x-array. However, the residuals must be generated manually by subtracting the predicted values from the test values as described hereafter.

Figure 2.8 shows that Excel has several nonlinear options for fitting data (linear, polynomial up to sixth order, logarithmic, exponential, moving average, etc.). When the linear option does not give a satisfactory fit, better results may be obtained with one of these.

Figure 2.10 shows the fit obtained using a second-order polynomial and Figure 2.11 shows the residuals. Unfortunately, Excel does not provide a route to generate residuals automatically except for linear fits using the tools method (see Section 2.2). However, it is a simple matter to generate and graph the required data. In this case, they are calculated on the sheet *polynomial* which can be found in the workbook *regress.xls*. The equation for the polynomial fit* given on the chart (see Figure 2.9) is used to generate the model data. For this particular case, the equation returned is

$$y = -0.00021x^2 + 0.0986x + 0.0224 \tag{2.2}$$

Instead of a simple subtraction of test and predicted data, the residuals have been normalized to the relative percent error by means of the equation

$$\% \text{ err} = \frac{(y_i - \hat{y}_i)}{y_i} \times 100 \tag{2.3}$$

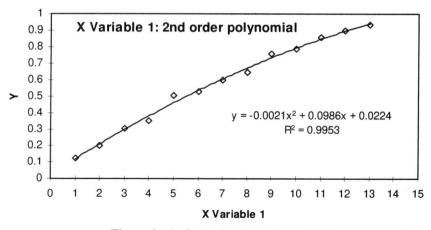

Figure 2.10. Second-order polynomial fit.

*Note that coefficients with insufficient precision may be obtained with higher order polynomials under the default settings in EXCEL. See Appendix A for further information.

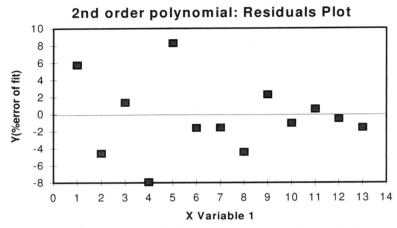

Figure 2.11. Residuals plot for second-order polynomial fit.

where

$$y_i = \text{the test datum for } x_i$$

$$\hat{y}_i = \text{the predicted value for } x_i$$

This type of transformation of residuals is useful for comparing the fit obtained with differently scaled data sets and gives an intuitive feel for the goodness of the fit. It also demonstrates the complete freedom the user has in arranging, scaling, or processing the worksheet data into a desired form. Figure 2.12 shows

	A	B	C	D
1	Polynomial Fit ($y = -0.0021x^2 + 0.0986x + 0.0224$)			
2				
3	**X-Index**	**TEST SET**	**Model Data**	**Residuals (%error)**
4	1	0.12617	0.1189	5.761794931
5	2	0.202	0.2112	-4.554455446
6	3	0.30349	0.2993	1.380481171
7	4	0.355	0.3832	-7.943661972
8	5	0.504704	0.4629	8.282897683
9	6	0.53	0.5384	-1.58490566
10	7	0.6004	0.6097	-1.548967355
11	8	0.648273	0.6768	-4.400519132
12	9	0.757006	0.7397	2.286166329

Figure 2.12. Calculation of residuals to Polynomial Fit—part of the sheet *polynomial* in the workbook *regress.xls*.

part of the data arrays generated with this particular data set, and Figure 2.13 gives the format used to input the corresponding equations for the model data in column C (equation 2.2) and normalized residuals in column D (equation 2.3). In both cases, the equations were entered in the first row (C4 and D4, respectively) and then filled down over the required range using the **Edit_Fill_Down** command. Hence, generating and graphing the data takes only a few moments (remember to use the CTRL key to enable nonadjacent columns to be selected for graphing (e.g., column A versus column D for the residuals plot; see Figure 2.12). The polynomial fit (Figure 2.10) is clearly better than the linear model, and the residuals (Figure 2.11) show a much more even scatter. Nevertheless, the residuals do show that the fit gradually improves as the x-value increases, suggesting that the second-order polynomial fit could be improved. Exercises like these are useful for reinforcing the importance of examining critically the fit obtained with a particular model and for emphasizing the importance of residuals in this process.

2.7. Inserting Error Bars

Perhaps the most common situation involving graphing scientific data is to generate a linear regression plot with y error bars. In most situations, the error in the x-data is regarded as being so much smaller than that of the y data, that, effectively, it can be ignored. Excel allows several methods for generating error bars, but we shall focus on one of these, *custom* error by which the experimental standard deviation of several estimations of a value can be used.

The workbook *regress.xls* contains five replicate y data arrays in the worksheet

	A	B	C	D
1				
2				
3	X-Index	TEST SET	Model Data	Residuals (%error)
4	1	0.12616963	=-0.0021*A4^2 + 0.0986*A4 + 0.0224	=100*(B4-C4)/B4
5	2	0.202	=-0.0021*A5^2 + 0.0986*A5 + 0.0224	=100*(B5-C5)/B5
6	3	0.30348961	=-0.0021*A6^2 + 0.0986*A6 + 0.0224	=100*(B6-C6)/B6
7	4	0.355	=-0.0021*A7^2 + 0.0986*A7 + 0.0224	=100*(B7-C7)/B7
8	5	0.50470412	=-0.0021*A8^2 + 0.0986*A8 + 0.0224	=100*(B8-C8)/B8
9	6	0.53	=-0.0021*A9^2 + 0.0986*A9 + 0.0224	=100*(B9-C9)/B9
10	7	0.6004	=-0.0021*A10^2 + 0.0986*A10 + 0.0224	=100*(B10-C10)/B10
11	8	0.64827263	=-0.0021*A11^2 + 0.0986*A11 + 0.0224	=100*(B11-C11)/B11
12	9	0.75700642	=-0.0021*A12^2 + 0.0986*A12 + 0.0224	=100*(B12-C12)/B12

Figure 2.13. The same sheet as in Figure 2.12 showing the formulas used to generate the values.

error bar that are to be graphed as the mean value with a linear regression fit to the mean, and with the standard deviations of the mean plotted as y error bars. Table 2.10 presents the data to be used in this exercise. Column A contains the data index (i.e., x-data series) and Column B contains the test data are the same as those used in the linear regression exercise in Section 2.6, but rescaled by a factor of 10). Column C contains the random error generated by means of the formula =**RAND( )*2**, which uses the Excel random number generator to produce the error. In Column D, this error is added to the test set by adding columns B and C. The next five columns (E to I) contain the simulated experimental data obtained by copying and pasting column D five times in succession. Each time the copy and paste actions are performed, the data in the random generator series are automatically updated, leading to a different error in each simulated data array. To obtain the five simulated data series;

- Highlight the data in column D.
- Copy to the Clipboard.
- Select the first cell of the column to which the data are to be pasted.
- Select **Edit+Paste_Special,** and select *Values* from the options.
- Click on OK to perform the transfer.
- The remaining cut and paste operations can be quickly completed by using

Table 2.10. Data for Generating Linear Regression Plot with Error Bars Shown in Figure 2.16

X-Index	Test Set	Random Error	Test Set + Error	$n = 1$	$n = 2$	$n = 3$	$n = 4$	$n = 5$	Mean $(n = 5)$	SD
1	1.262	0.557	1.819	2.419	2.004	1.283	3.128	2.135	2.194	0.669
2	2.020	1.434	3.454	2.350	3.007	3.497	2.823	2.135	2.762	0.540
3	3.035	0.431	3.466	4.729	3.163	4.623	4.699	3.803	4.203	0.697
4	3.550	0.259	3.809	5.094	3.851	5.194	4.599	3.732	4.494	0.681
5	5.047	1.295	6.342	5.810	5.090	6.903	5.266	6.665	5.947	0.813
6	5.300	1.511	6.811	5.875	6.230	6.832	6.739	6.838	6.503	0.432
7	6.004	1.005	7.009	6.077	7.780	7.978	7.142	7.082	7.212	0.745
8	6.483	0.742	7.224	6.826	6.766	7.817	6.517	8.269	7.239	0.760
9	7.570	0.792	8.362	8.294	8.097	8.515	9.134	9.559	8.720	0.610
10	7.903	0.424	8.326	9.463	8.992	8.352	9.597	9.119	9.105	0.487
11	8.582	1.987	10.569	9.368	8.974	8.693	9.148	9.371	9.111	0.287
12	8.986	0.738	9.724	10.484	10.269	10.473	10.635	10.421	10.456	0.132
13	9.348	0.221	9.570	11.282	10.859	9.846	9.772	9.553	10.262	0.760

the **Edit+Repeat_Paste** command, and selecting the top cell of each succeeding column (see Figure 2.14, columns E–I).

- The mean and standard deviations are obtained from the average() and stdev() commands in the last two columns (see Figure 2.14, columns J and K).

The objective is to plot the mean and obtain the regression line to the averaged data in column J and then to add error bars given by the standard deviations in column K (Figure 2.14).

- Generate a scattergraph of the x-data index vs. the mean (use the left mouse button and the CTRL and SHFT keys to highlight the two columns) and select the option that plots unjoined points when generating the graph (i.e., no line).
- Use the Insert Trendline option to generate a linear fit to the mean as described.
- Highlight the data in the graph window (single click mouse on any of the data points).
- Select the command **Insert_Error Bars;**
- Choose **y error bars** option (note that an x error bar option is also available,

	E	F	G	H	I	J	K
1							
2		data set					
3	n=1	n=2	n=3	n=4	n=5	mean (n=5)	SD
4	2.41866	2.003888	1.28291	3.12787	2.13497	=AVERAGE(E4:I4)	=STDEV(E4:I4)
5	2.34983	3.007242	3.49695	2.82269	2.13509	=AVERAGE(E5:I5)	=STDEV(E5:I5)
6	4.72872	3.162773	4.62345	4.69935	3.80281	=AVERAGE(E6:I6)	=STDEV(E6:I6)
7	5.09448	3.851152	5.19397	4.59936	3.73197	=AVERAGE(E7:I7)	=STDEV(E7:I7)
8	5.80991	5.090469	6.90256	5.26566	6.66470	=AVERAGE(E8:I8)	=STDEV(E8:I8)
9	5.87546	6.230185	6.83238	6.73882	6.83840	=AVERAGE(E9:I9)	=STDEV(E9:I9)
10	6.07730	7.779975	7.97843	7.14193	7.08182	=AVERAGE(E10:I10)	=STDEV(E10:I10)

Figure 2.14. Formulas used to calculate mean and standard deviation of electrode relaxation potentials for five runs.

Note: It is important to select the Values option when performing the paste as this transfers only the values generated by the formulas in the copied cells and not the formulas themselves. This is useful when using the random number generator, as this function keeps updating the cell values each time a spreadsheet operation is performed, and tends to slow worksheet operations generally, particularly if a graphic window is linked to the cells being updated.

if this error is known); a window similar to that shown in Figure 2.15 will appear.

- Select **both** under the Display options and **Custom** under the Error Amount options.
- The Custom option requires the user to input the error source for each point in the y-data array. To do this, click the + field to the right of the Custom button, move the mouse to the spreadsheet and click once to activate the sheet. Move the pointer to the first cell in the standard deviation array (K4) and click the mouse button. Drag the mouse to the last cell in the array, or alternatively, move the pointer to the last cell in the array, and click while holding the SHIFT key down. The array addresses can also be entered manually using a colon to signify an array in the normal fashion (K4:K18).
- Move the pointer back to the error bar window and activate it by clicking.
- Select the − error field below the + field and enter the standard deviation array address as above (the standard deviation is symmetrical about the mean).
- Click on OK to complete the operation.

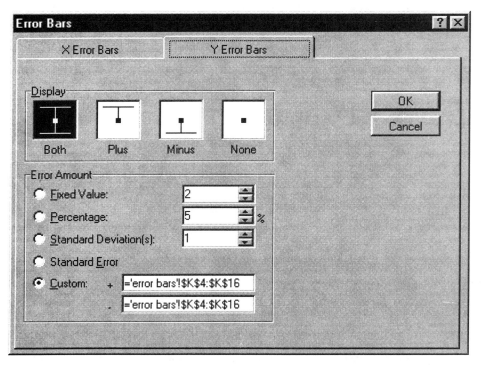

Figure 2.15. Dialog box for Insert_Error Bars.

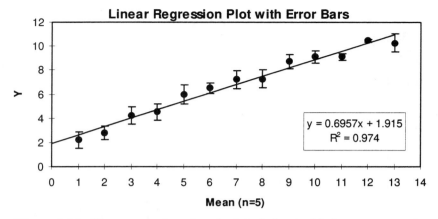

Figure 2.16. Custom error bars (standard deviations) added to scatter graph.

The final product should be similar to Figure 2.16. Many useful exercises can be generated from the data available in this workbook. For example, the five simulated data sets in the sheet *error bars* could be copied to a new sheet and the task of generating the graph in Figure 2.16 set. Linear regression and nonlinear regression tasks can be set using the data in the sheets *linear regression* and *polynomial*. A more detailed case study is presented in the next section, which involves interpreting experimental data with a nonlinear (exponential) model.

2.8. Non-linear Regression—Modeling Potential Relaxation in Membrane Electrodes

The sheet *electrode data* in the workbook *regress.xls* contains two sets of data obtained from experiments in which the relaxation of the membrane potential of two electrode types, A and B, was monitored after a step increase in the membrane potential was caused by injecting ions to which the membranes respond. The experiment was repeated five times for each electrode.

- Obtain the average and standard deviation of each point for both electrodes.
- Plot scatter graphs of these (without joining points).
- Fit an exponential regression model to the data.
- Add y error bars using the Custom error option as described above. The resulting fits are shown in Figure 2.17 (type A) and Figure 2.18 (type B).

From these graphs we can see that

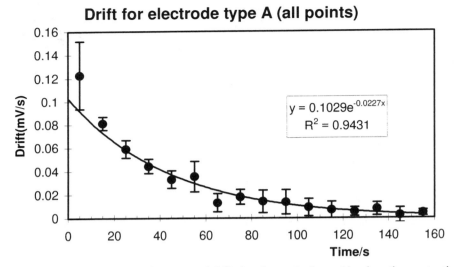

Figure 2.17. Error bars and exponential fit for electrode (type A) relaxation potentials.

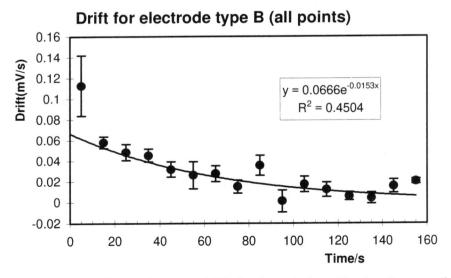

Figure 2.18. Error bars and exponential fit for electrode (type B) relaxation potentials.

- The exponential model appears to fit the data for electrode A better than type B.
- The first and second points seem to be much higher than the rest of the data.
- The standard deviation of the measurements generally improve with time.

At this point, the experimenter must decide if the single exponential model describes the relaxation process adequately or whether a more sophisticated model is required. More complex models are explored in detail in chapter 7 using *Solver*. A relatively simple option is to delete the first point from the data set and remodel using the single exponential fit. This can be justified on the basis that the initial point in an exponential series of experimental data is often the most inaccurate. The size of the error bars for these points is clearly much larger than for the rest of the data, and the points are well outside the trend. In addition, the relaxation process follows a very fast initial step increase in potential (not shown), and the two processes are certainly merged to an unknown degree. If these arguments are accepted, then the initial point could be causing a significant skew in the model and should be deleted. Figure 2.19 shows the combined models for both electrode types, with the initial point deleted in each case. This figure suggests that there is little difference in the relaxation behavior of the two electrode types, and both have single exponential half-lives of around 30 to 35 s.

Once again, from the case study in this section, it is clear that modeling the

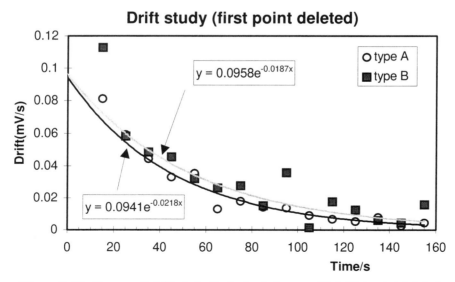

Figure 2.19. Exponential fits to both electrode types, with initial point deleted.

data is part of a larger process that requires the experimenter to possess significant background knowledge of the subject area in order to justify both the modeling strategy (which model to use, whether to delete certain points, etc.) And the values returned by the model. And, as often is the case in science, there may be more than one modeling approach and alternative interpretations of the results, which in the end is what makes science interesting!

REFERENCE

[1] Miller, J. C. and Miller, J. N. 2nd ed. *Statistics for Analytical Chemistry*. Ellis Horwood Ltd., Chichester, UK, 1988.

CHAPTER 3

INTRODUCTION TO MACROS AND VBA

3.1. MACROS

A macro is a sequence of Excel commands grouped together and run as a single command. Excel versions 5 and 7 allow for backward compatibility with Excel version 4 and therefore support two macro languages, Excel 4 macro language and Visual Basic for Applications (VBA). However, as Microsoft has decided to move with VBA as its primary macro language, it is a good idea to curtail programming in Excel 4 and make the plunge into VBA. This book deals only with VBA macros. Keep in mind that when one refers to a macro in Excel it may be an Excel 4 macro or a VBA macro.

The commands in a macro are related in the sense that they accomplish a larger task. Macros are a method for accomplishing the following.

- Automating frequently used sets of commands.
- Automating complicated tasks.
- Reducing the number of steps in a complex operation.
- Making complex formula bar entries more efficient.

All the icon buttons in the toolbars are associated to macros. When the button is depressed, a macro is invoked. The result is a combination of commands from the drop-down menus. For example, the PrintPreview icon is a macro which runs File-Print Preview and the Bullets icon is Format_Bullets and Numbering_Bulleting. Wizards are also macros. Macros can be simple or very involved depending on the task, energy, and enthusiasm of the programmer. Since macros are collections of a series of commands, once they are run, *Undo* cannot reverse them.

3.1.1. Formatting Macro

A quick overview of macros is best understood by *recording* one. Most macros are not written but are recorded mouse actions. The following case study is to record a macro to format the cells of a title row on a spreadsheet. This macro will be recorded twice, first with the range A2:C2 highlighted and then with the range not highlighted before recording. The differences of their execution will be commented on afterwards.

Case Study 3.1. Recording a Macro

Record a macro to format the cells of a title row of a spreadsheet with the range highlighted prior to recording.

- Open a new workbook and name it *chap3.xls* by saving it.
- Double click the Sheet1 tab and give it the name *format*.
- Enter in the titles **Theoretical Values, Experimental Values**, and **% Deviation** in cells A2 to C2.
- Select **Tools_Record Macro_Record New Macro**.

A dialog box (Figure 3.1) comes up for the name of the macro to be entered (*Macro1* is default name).

A macro name cannot have spaces or punctuation marks, with the exception of the underscore, _. The macro name needs to reflect the action of the macro.

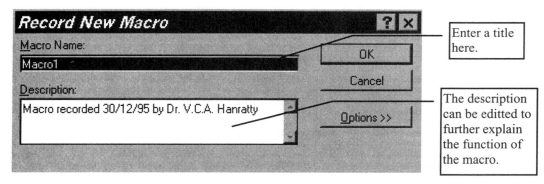

Figure 3.1. The Record New Macro dialog box. Excel defaults the name *Macro*. The Options button allows further customizing of the macro.

In this case study, the name used is FrmtCells1. Choose a different name if you desire.

- Use the TAB key to move to the description box for altering and enter in something to the effect that A2:C2 are highlighted before recording.

Click the **Options** button and have a look at the options that can be set. This dialog box determines what the macro can be assigned to, where to store it, and in what language to record it. It is from here that a macro can be assigned to a keypress or menu item. See Figure 3.2.

Ensure that the options are set to store the macro in **This Workbook** and the language is **Visual Basic**. Pressing CANCEL will escape out of recording the macro.

- Press OK when finished.
- A small Stopbox ▣ appears on the spreadsheet after the Macro dialog box closes. Its inset black square button will be clicked when recording the steps of the macro is complete. Every mouse action now taken will be recorded in the macro.

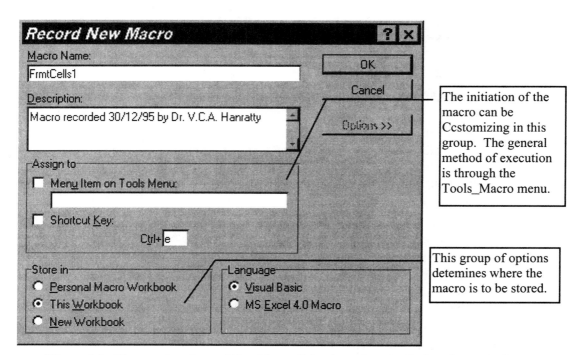

Figure 3.2. The complete Record New Macro dialog box when the Options button is chosen.

If a few inadvertent mouse actions are taken along the way, don't worry; their effect on the overall task of the macro may not be detrimental and they can be edited out after recording is complete.

Record the macro by carrying out the following mouse actions.

- Select the bold icon **B** and the italic icon *I*.
- Change the font size to **12** by clicking the drop-down Font Size box and click 12.
- Select **Format_Column_AutoFit Selection** to adjust column widths automatically.
- Stop recording by pressing the inset black square button in the Stopbox.

The macro is finished and its code has been listed in a *module sheet*. Scroll to the end of the sheets and find the tab labeled module1. This is a *VBA macro sheet*. The name *macro sheet* has been retained for Excel 4 macros, hence Excel 4 macros are recorded on a macro sheet and VBA macros on a module sheet. Double click the module1 tab and enter the name **Frmt Macro**. Do not be alarmed in viewing the contents of this sheet when its tab is selected. This sheet contains the FrmtCells1 macro written in VBA. Knowledge of VBA is not necessary to record and run macros. However, you should be able to recognize the tasks you asked the macro to do within the code. The code will be illustrated in Section 3.2.

Case Study 3.2. Running a Macro

The FrmtCells1 macro is run to ensure that it works.

- Scroll back to the *format* spreadsheet and select it. Since in recording the macro the cells of row 2 have already been formatted, their contents need to be cleared.
- Highlight the title cells and choose **Edit_Clear_All**. This clears contents and formatting.
- Select A2:C2 and change the font back to size **8** through the Font Size drop-down box.
- Readjust the column widths back to the standard width (8.43) by highlighting the column headings and choosing **Format_Column_Standard Width**. An alternative method is to drag the column divider to the desired width, though this method can be awkward for some. The value of the column width can be read in the Name box.
- Reenter the titles in A2:C2.
- Make sure the range is highlighted before running the macro.

- Select **Tools_Macro**. From the Macro dialog box, select your macro from the list of existing recorded macros and press **Run**. This macro may be the only macro listed.

The titles should be formatted accordingly. If the macro does not work, it is best to record it again by giving the same macro name (clear the cells and adjust columns again). The old FrmtCells1 will be overwritten. You might first like to compare your code with Table 3.1 and see if you can identify the problem.

Case Study 3.3.

Record a second macro to format the cells of a title row of a spreadsheet with the range not highlighted prior to recording.

- For the range A2:C2, select **Edit_Clear_All** and change the font back to size **8**.
- Reenter the titles.
- Record a new macro with the name **FrmtCells2** and enter a description that the A2:C2 range is selected by the macro.
- The same mouse actions as for FrmtCells1 are used with the insertion of a first step:
  - Highlight the range A2:C2.
  - Select the bold icon **B** and the italic icon *I*.
  - Change the font size to **12** by clicking the drop-down Font Size box and click 12.
  - Select **Format_Column_AutoFit Selection** to adjust to column widths.
  - Stop recording by pressing the inset black square button in the Stop-box.

This macro is recorded on the same module sheet as FrmtCells1. Macros recorded in the same Excel session are usually placed in the same module sheet. The two macros differ only in terms of whether or not the intended range is selected before recording. Therefore, when one or the other macro is run, the selecting action must be taken accordingly. When running FrmtCells1, the selection must be highlighted beforehand. With FrmtCells2, this action is not necessary, as the intended range to be formatted, A2:C2, is selected as part of the macro. There is more flexibility in using FrmtCells1. This macro will format any range on the spreadsheet, as the intended range is selected prior to running the macro.

Case Study 3.4.

Visualizing the Effects of Running FrmtCells1 without Prior Selection of the Title Range

- Edit the spreadsheet back to square one and reenter the titles in A2:C2.
- Do not highlight the range and run the FrmtCells1 macro.
- The title cells were not formatted. Instead, the row containing whatever cell was active prior to running the macro was formatted. Even though the cell may have been empty, the font size has been changed and you will notice that this row is of different height than the rest.

It is generally recommended to highlight cells before recording macros, as this usually makes the macro more portable within and between spreadsheets. However, this course of action will depend more on the task of the macro than following the generality. It is necessary, though, to be able to recognize the difference in VBA code of the two format macros because a common edit in macros is to change the selection action on cells from within the code to use a selection made before running the macro or vice versa. This will be demonstrated in the following section. In practice, it has been found that to select cells before recording a macro involving formatting and for macros performing calculations, it is better programming practice to make the selecting part of the macro. In this way you know for sure the numbers are going where they are supposed to go.

3.1.2. Calculation Macro

This next macro recording will perform calculations on data. This macro will calculate the average result of three experiments. The first task is to create the data representing results of three experiments.

Case Study 3.5.

Create a spreadsheet that will represent the results of three repeated experiments of analysis. The techniques for entering data are reinforced, for example, Edit_Fill_Series and AutoFill.

- Start with a clear sheet by clicking the next sheet tab.
- Name this spreadsheet *calcdata*.
- In cells A2:C2 put in the titles **Run1, Run2**, and **Run3**, respectively. Enter them by AutoFill. Enter **Run1** in A2, press ENTER and then reclick A2 to make it active. Drag its fill handle across to C2 and release. Run2 and Run3 have been automatically entered by AutoFill.

- Put **Average** in D2.
- Use **Edit_Fill_Series (linear)** to enter data into cells A3:A12. Start by entering **5** in A3. Highlight the range A3:A12, select Edit_Fill_Series, choosing **column, linear**, and **step value 1.03**.

The data for B3:C12 will be entered using Edit_Fill_Series (trend), but first a minimum of data needs to be entered in random cells of this range for the trend to work.

- Type in about four entries in each column (B and C) either slightly less or greater than the corresponding value in that row of column A. An example of this is in Figure 3.3.
- Highlight B3:C12.
- Bring up the series dialog box and tick **columns** and the **trend**.

Excel has completed the data in columns B and C by determining a trend based on the data present. Notice that the original entries have been altered. This is to give a smooth trend through the preselected range. The resulting data from the trend calculation is to the full precision of Excel (15 decimal places).

- To make all the data consistent in decimal places, select the A3:C12 range and click the decrease decimal icon of the formatting toolbar until all the data have two decimal places.

	A	B	C	D
1				
2	Run1	Run2	Run3	Average
3	5.00	4.9		
4	6.03		6.1	
5	7.06			
6	8.09	8.1		
7	9.12		9.0	
8	10.15	9.9		
9	11.18		11.2	
10	12.21			
11	13.24		13.4	
12	14.27	15.0		

Figure 3.3. The spreadsheet in preparation for data entry by the Fill_Series command filling the cells according to the present trend.

Two macros will be recorded to calculate the average, to examine how to assign the use of relative and absolute references in macros. It is advantageous to use relative references to allow portability of your code, but which you choose most likely depends on your own preference, the particular spreadsheet, and the task. In both macros, the intended cell D3 (where the macro is to begin) will be highlighted before recording is begun.

Case Study 3.6.

Recording a Macro to Calculate the Average on a Set of Data using Absolute References

- Highlight D3.
- Start the recorder (Tools_Record Macro_Record New Macro). Name the macro and the description, using absolute references in VBA code. The macro name used in this task is CalcAvg1.
- Press the **Options** button and ensure the macro will be stored in **This Workbook** and in VBA. Press OK when finished.

Macro Mouse Actions

- Click in the formula bar and enter =**AVERAGE(A3:C3)**. [An alternative method of entering references in a formula is to select the cells on the spreadsheet. After typing in =**AVERAGE(**, highlight the range A3:C3 on the spreadsheet. A dotted line (hatched line) box should be outlining this range. In the formula bar, A3:C3 appears; press RETURN.]
- Highlight D3:D12 and select **Edit_Fill_Down** or use the shortcut key-press of CTRL+D.
- Stop the recorder.

The code for this will most likely be found on the *frmt macro* sheet if it was recorded in the same workbook and session as the Frmt macros. It will be in a newly created module sheet if this the first macro recorded since opening the workbook.

- Scroll to the end of the sheets and find in which module sheet CalcAvg1 was entered.
- If it is in a new module, give this module the name **Avg Macro**.

If it is in the *frmt macro* sheet, cut and paste it into a new module sheet. This is done through Insert_Macro_Module.

- Select **Insert_Macro_Module** and give it the name **Avg Macro**. (Right clicking on any sheet tab brings up the shortcut menu where Insert-Module can be chosen.)
- Highlight the complete code of CalcAvg1 and **cut**. This entry starts with the comment lines naming the macro, those lines beginning with a single quote, to **End Sub**.
- Select the *Avg Macro* tab and **paste**.

Case Study 3.7.

Record the second macro to calculate the average on a set of data, this time using relative references.

- Scroll to and select the *calcdata* spreadsheet.
- Clear the contents of the cells D3:D12.
- Select D3. This cell should now be empty.
- Record the macro for using relative references by selecting **Tools_Record Macro_Use Relative References**. Calling up Tools_Record Macro again will show a checkmark beside Use Relative References. This time choose **Record New Macro**.
- Give the macro name **CalcAvg2** and the description, **using relative references in VBA code**.
- Record the exact same mouse actions as in Case Study 3.6 and stop the recorder when finished.

Scroll to the module sheets and find where CalcAvg2 was placed. If this macro was placed in the *frmt macro* sheet, cut and paste it on the *avg macro* sheet.

Case Study 3.8. Comparison of CalcAvg1 and CalcAvg2

Examine how the recorded macros using absolute and relative references differ in operation when the data is moved to a new location on the spreadsheet.

- Select the *calcdata* spreadsheet.
- Clear the contents of D3:D12.
- Highlight the current region of the spreadsheet, that is, A2:D12.
- Move the contents of these cells to a new location on the spreadsheet by placing the mouse on the edge of the outlined range until a left pointing arrow appears. Now drag the box to the right. For example, move the right

corner of the selected region to F2. Release the left mouse button. See Figure 3.4.

- Select the cell immediately under the Average title cell (I3).
- Run the CalcAvg1 macro.

Were the averages calculated and placed under the Average cell? No, they were not, because this macro was recorded using absolute references. The destination of the calculated values was determined during recording as D3:D12, not I3:I12 where the data are presently. The first entry in the Average column is correct because this entry was evaluated from the entered formula where relative references were used. After this action is completed, the macro highlighted D3:D12 to Fill_Down, but since D3 was empty, this range is blank.

- Clear the contents of the cell immediately under the Average title cell and reselect it.
- Run the CalcAvg2 macro.

Did the macro work as intended? Where were the calculated averages placed? They should have been placed in the Average column (I3:I12). This macro was recorded using relative references, meaning no matter where the

Current Region, A2:D12.

Move current region by dragging area to F2.

	A	B	C	D	E	F	G	H	I	
1										
2	Run1	Run2	Run3	Average						
3	5.00	4.79	5.07	4.95						
4	6.03	5.90	6.07	6.00						
5	7.06	7.02	7.08	7.05						
6	8.09	8.13	8.09	8.10						
7	9.12	9.24	9.09	9.15						
8	10.15	10.36	10.10	10.20						
9	11.18	11.47	11.11	11.25						
10	12.21	12.58	12.11	12.30						
11	13.24	13.70	13.12	13.35						
12	14.27	14.81	14.13	14.40						
13										

Figure 3.4. New location of data. Move right corner of highlighted area to F2 as shown in the figure.

data is placed on the spreadsheet, the calculated values are placed relative to the data. This is provided, of course, that the formula entered during recording is done with relative references.

Macros recorded using absolute references use the absolute location of cells in determining destination. Those using relative references decide the locations of cells relative to the positions of previously selected cells. In CalcAvg1, the final destination was the exact range of D3:D12. D3 was empty so D3:D12 was filled as empty. In CalcAvg2, the final destination was determined based on the relative position of the data to the selected cell. Use of relative references allows portability of the data to any location on the spreadsheet and the calculation results are placed in the cells relative to the present location of the data.

3.2. Introduction to Visual Basic for Applications

Visual Basic for Applications (VBA) is the programming language for Excel macros. VBA is essentially a subset of the Visual Basic (VB) programming language. It is presently shipped with Excel and is soon to be included in all the office applications to give a consistent macro language. It is a user-friendly language, also known as an event language, and strives to automate the programming process. VBA has succeeded in making programming an approachable and achievable undertaking, especially to those who do not visualize themselves as programmers or program developers. It makes the learning process bearable because the learning curve is short, and gives courage to the weakhearted as programming accomplishments are rapidly achieved.

VB is the newest generation of the Basic programming language. Its trademark is introducing event structures to make programming and the running of programs a visually simple task. It has taken the *writing* out of programming, replacing it with recording tasks. VBA is an object-oriented programming language. In Excel, the objects are the application, workbook, worksheet, range, and others, with the addition of useful VB objects. Programming in VBA is controlling the objects of Excel to perform desired tasks. VBA controls objects with their properties and performs action with the object's methods. *Object, property*, and *method* are important terms in VBA. In essence they are the basis of VBA syntax. Though this syntax may be new, don't let it deter you. Just remember that methods and properties are ways to work with objects and you'll pick up the syntax as you view your recorded macro code. You can read more on objects, methods, and properties by making active a module sheet and selecting View_Object Browser. Figure 3.5 shows the Excel object library.

From the Libraries/Workbooks drop-down box, the VBA library can be chosen to view its objects and corresponding methods and properties. For further details on VBA, the reader is referred to On-Line Help and the *Microsoft Excel Visual Basic User's Guide.*

By far the easiest way to learn VBA is to follow the code for a task you understand and learn how VBA performs this task. We will begin our introduction to the VBA language by reviewing the code created when the macros were recorded in the previous section: FrmtCells1, FrmatCells2, CalcAvg1, and CalcAvg2. The structures and syntax of their code will be simply described line by line. This will give a working knowledge of VBA, its commands and vocabulary.

3.2.1. Code of the Formatting Macros

We will start with the first two macros recorded, FrmtCells1 and FrmtCells2. Select the *frmt macro* sheet and scroll down the sheet to ensure that the two macros, FrmtCells1 and FrmtCells2, are present. The contents should be something similar to that found in Table 3.1. If your macro contains additional lines, it is most likely the result of extraneous mousing around. Don't worry about them unless your macro did not format as intended, in which case rerecord the macro.

Notice that the first lines of the macro begin with a single apostrophe mark, '. This apostrophe denotes the line as a comment line as opposed to a line of exe-

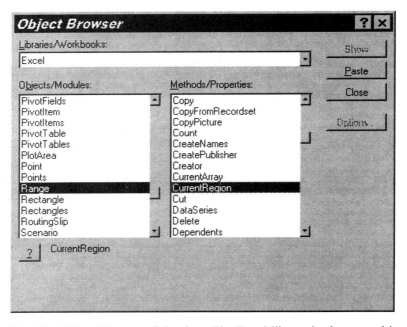

Figure 3.5. The Object Browser dialog box. The Excel library is chosen and its objects displayed. When an object is selected its corresponding methods and properties are shown.

cutable VBA code. Comment lines are important enhancements to remark on the particulars of the macro and explain the purpose of sections of code. The first few comment lines should be the macro name and description entered in the Record New Macro dialog box. These lines are usually preceded and followed by a couple of blank comment lines. The VBA displays portions of code in various colors for easy identification of their purpose. The default colors are green for comment lines, blue for command keywords, black for statements, and red for syntax error code. These colors can be changed in the Tools_Options_Module Format (tab) dialog box.

The **Sub** line is the first executable line of code in the macro and all macros. Sub is an abbreviation for subroutine, a type of procedure in VBA. In essence a macro is a sub. A macro can be comprised of many subs or be just one sub. The macro finishes with the **End Sub** line. All lines of code between the Sub and End Sub lines comprise the macro.

Let us review the terms *module, macro, procedure, sub*, and *function* used in

Table 3.1 VBA Code for the Formatting Macros, FrmtCells1 and FrmtCells2, as Recorded[a]

```
' FrmtCells1 macro
' A2:C2 highlighted before recording
'
Sub FrmtCells1()
   Selection.Font.Bold = True
   Selection.Font.Italic = True
   With Selection.Font
      .Name = "Arial"
      .FontStyle = "Bold Italic"
      .Size = 12
      .Strikethrough = False
      .Superscript = False
      .Subscript = False
      .OutlineFont = False
      .Shadow = False
      .Underline = xlNone
      .ColorIndex = xlAutomatic
   End With
   Selection.EntireColumn.AutoFit
End Sub
```

```
' FrmtCells2 macro
' A2:C2 not highlighted before recording
' that action is the first step of the macro
'
Sub FrmtCells2()
   Range("A2:C2").Select
   Selection.Font.Bold = True
   Selection.Font.Italic = True
   With Selection.Font
      .Name = "Arial"
      .FontStyle = "Bold Italic"
      .Size = 12
      .Strikethrough = False
      .Superscript = False
      .Subscript = False
      .OutlineFont = False
      .Shadow = False
      .Underline = xlNone
      .ColorIndex = xlAutomatic
   End With
   Selection.EntireColumn.AutoFit
End Sub
```

[a]Compare the code, keeping in mind that the recorded mouse actions differ only in when the title cells were selected.

VBA. You may already be familiar with these terms from other programming languages. In VBA, a module sheet is an Excel sheet where VBA code is entered. A module contains *procedures*. A procedure is a unit of code entered in the VBA module and is executed as a whole to perform a specified task. There are two types of procedures in VBA, *subroutines* and *functions*. They differ fundamentally in what is known as a *return value*, though both accomplish a task. Subroutines have no return value and functions return a value. Subroutines generally take no arguments and a function has arguments passed to it and returns a value to a subroutine or spreadsheet. Arguments are used when values for variables are being passed from one procedure to another. The use of function procedures by spreadsheet and an example of a function called by a subroutine are discussed in Section 3.3. The macros recorded in the previous section are all sub procedures and their tasks have been either formatting or calculation.

The term *macro* is used freely in Excel. It can refer to both Excel 4 and VBA macros. It has already been defined as a sequence of commands grouped together and run as a whole. This sequence of commands are the subs and functions which reside in a module sheet, so when one refers to a macro, it can be a sub, a few subs, subs and functions, or the entire contents of the module sheet. *Macro sheet* and *module sheet* are used interchangeably in this book, though in both instances they refer to VBA code.

To continue with the code description, the FrmtCells1 macro is defined by the lines between the Sub and End Sub lines.

- The syntax for the Sub statement is **Sub** *Name* (argument list). Notice there are empty parentheses in the Sub line of FrmtCells1, meaning there are no arguments in this sub. The function procedures usually have arguments.
- The next two lines in the FrmtCells macro are recognized as changing the font to bold and italic. *Selection* refers to the active cell (or cells) at the time of recording the macro. And this is just how FrmtCells1 was recorded; the range A2:C2 was highlighted and then the recorder was started. The font is edited in the Selection lines.
- The next series of lines are grouped by *With* and *End With* lines. The With statement is a method to perform a series of actions. In this case, the series of actions is on the font. All the lines contained in this With statement are those of Format_Cells_Font dialog box. Take a look at this dialog box and compare the selections with the contents of the With statement, recalling the mouse actions performed in recording this macro; selecting the Font Size box on the formatting toolbar created this With statement. So you can see how VBA produced code from selections made in a dialog box. It placed the complete set of choices for font in the macro from the primary drop-down menu.

- The last *Selection* line, before the End Sub, performs the AutoFit on the columns.

Consider for a moment the effect of changing the sequence of steps of Frmt-Cells1. What would be the result if autofitting the column widths was done before changing the font to bold, italic, and size 12? If in doubt, copy the AutoFit line (*Selection.EntireColumn.AutoFit*) to the beginning of the macro, just after the Sub line, and run again on a clear sheet. The result is that the columns are too narrow for the edited text. What if the titles were entered after the macro was run? Again the columns are not wide enough to show the text entries. As you can see, consideration must be given to the sequence of steps in macros. In this example, their order makes no consequence to the error-free running of the macro, though it does to the overall effect on the task.

3.2.2. Editing Macros

Recording macros does not always produce the most efficient code. Instead, many macros result in repetitive and redundant lines of code, though the task is still performed. In simple macros, editing of the code is not required; it would be an unnecessary effort. However, developing larger macros for more complicated tasks usually requires the good programming practice of cleaning up the code, that is, editing out unnecessary lines. In the code listings of Table 3.1 there are many assignment statements to *false* as well as two sets to *none* and *automatic*. These are examples of redundant lines of code. In the FrmtCells1 macro there are seven such lines and these can be safely deleted without altering the action of the macro. Go ahead and delete these lines now.

Perform further tidying on the repetitive actions in FrmtCells1 that you may have noticed. The font has been set to bold, italic in the first two Selection lines and also in the With statement. Therefore, delete lines 2 and 3, the Selection lines. After these edits, the code for the FrmtCells1 macro should look like Table 3.2. The code is now more efficient. It contains only the necessary lines of code to carry out the desired task and actions are not being repeated. Practice these editing skills by tidying the code for the FrmtCells2 macro. The resulting code should be as shown in Table 3.3.

In Section 3.2.1, these two macros were compared in how they operated in formatting a range. FrmtCells1 required that the desired range to be formatted be selected prior to running and FrmtCells2 did not. Viewing the code shows how VBA denotes this difference. The two macros are the same with the exception of the extra line at the beginning in FrmtCells2 Range("A2:C2").Select. This demonstrates the code for selecting a range as the object of further code. In FrmtCells1, the range object is selected prior to running the macro; therefore, the object of the With statement is Selection.

Table 3.2. Edited Code of the FrmtCells1 Macro

```
Sub FrmtCells1 ()
 With Selection.Font
   .Name = "Arial"
   .FontStyle = "Bold Italic"
   .Size = 12
 End With
 Selection.EntireColumn.AutoFit
End Sub
```

3.2.3. Code of the Calculation Macros

Turning to the CalcAvg macros, select their sheet tabs and compare your codes with those in Table 3.4 and Table 3.5. These two macros differ in that one was recorded using absolute references (CalcAvg1) and the other using relative references (CalcAvg2). The relative and absolute references referred to here are in relation to that used in VBA code. This is different from the referencing used in formulas entered in the spreadsheet that was discussed in Section 1.2. View the code of the two macros and try to identify where this difference is demonstrated. The destination cells for the Fill_Down operation have been determined by absolute references in CalcAvg1 by the code Range("D3:D12").Select and by relative references in CalcAvg2 by ActiveCell.Range("A1:A10").Select.

The absolute reference of the Select action, directing the results of the Fill_Down operation, is denoted in the code by the keyword **Range**, naming specific cells, D3:D12. Conversely, the relative reference is denoted by *ActiveCell.Range*, naming the destination for the Fill_Down task relative to the present active cell. In this case, the active cell is the cell where the formula was entered, relatively referred to as A1, and the fill starts from that cell downward by A1:A10. Hence, this is how relative references make macros portable. They are not attached to specific cell addresses.

Table 3.3. Edited Code of the FrmtCells2 Macro

```
Sub FrmtCells2 ()
        Range("A2:C2").Select
        With Selection.Font
                .Name = "Arial"
                .FontStyle = "Bold Italic"
                .Size = 12
        End With
        Selection.EntireColumn.AutoFit
End Sub
```

Table 3.4. Code for the CalcAvg1 Macro

```
' CalcAvg1 Macro
'  ABSOLUTE REFERENCES BEING USED
'
Sub CalcAvg1()
    ActiveCell.FormulaR1C1 = "=AVERAGE(RC[-3]:RC[-1])"
    Range("D3:D12").Select
    Selection.FillDown
End Sub
```

The first line in both macros has the average formula entered in the active cell of the spreadsheet with relative references. As this was done in macro record mode, VBA used R1C1 type. Excel defaults to relative cell references. Remember, it is easy to change the referencing while in the formula bar with the F4 key. VBA records the referencing type you choose.

There is a VBA toolbar that is useful to have displayed when working on a worksheet developing macros. This shortcuts having to go into the drop-down menus to select commands. When in a macro sheet, the VBA toolbar is displayed by default. You may have noticed it before when you were in View_Toolbars. Display the VBA toolbar now from this menu and move the mouse arrow over each button to familiarize yourself with its function from the text description that automatically appears. A handy button is the green arrow for Run Macro since this action is used repeatedly when constructing a macro.

A very useful macro to record is that to customize the formatting of an XY Scatter graph, as this is the most common type of graph used in analyzing scientific data. The preferences of style for plot area, trendline line type, and so on can be recorded on an existing graph. These would be the formatting that you do each time on a graph. This macro can then be used on any graph to allow consistent formatting on all graphs you create. Assigning this macro (or any macro) to a key press allows it to be activated quickly. To do this choose Tools_Macro, select the

Table 3.5. Code for the CalcAvg2 Macro

```
' CalcAvg2 Macro
'  RELATIVE REFERENCES BEING USED
'
Sub CalcAvg2()
    ActiveCell.FormulaR1C1 = "=AVERAGE(RC[-3]:RC[-1])"
    ActiveCell.Range("A1:A10").Select
    Selection.FillDown
End Sub
```

macro you wish to assign to a key press, and choose the Options button. This is the same dialog box seen when a macro is first recorded. It is in this dialog box that the macro can be assigned to a keypress of your choosing. Try not to override any built-in keypresses. It would also be very helpful to have a macro of this type stored in the *personal.xls*.

3.3. FUNCTIONS

As stated earlier in this section, a function uses arguments and returns a value. Functions will be demonstrated first through the writing of a user-defined or custom function. Built-in Excel functions were discussed in Section 1.3 through describing the use of the Function Wizard tool. However, Excel allows the user to introduce new functions through programming. A user-defined function is valid on any spreadsheet within the workbook where it was stored. If the function is saved to the *personal.xls,* then it can be used in any workbook or globally. The *personal.xls* serves as a global retainer for macros, as well as other Excel features.

Recalling the percentage calculation used in the *iserror* spreadsheet, the *iserror* sheet of *chap1.xls* will be revisited for the custom function demonstration. This spreadsheet calculated the percent mass of C, H, and O in ethanol. A function can be written to calculate the percent, which can then be used in the same manner as Excel's built-in functions. Simply, the formula in the cell of the spreadsheet will be replaced with a custom function.

Case Study 3.9. Writing a Custom Function to Calculate a Percentage for Use in the Local Workbook

- Copy the *iserror* sheet from *chap1.xls* to *chap3.xls*. Open *chap1.xls*, select the IsError sheet tab and choose **Edit_Move** or **Copy Sheet**. Remember to select the Create a Copy check box.
- Select the IsError sheet tab and choose **Insert_Macro_Module** to bring in a new module sheet.
- Select the module sheet tab and give it an appropriate name.
- Type in a few comment lines: lines that begin with an apostrophe or single quote, ', describing the name and purpose of the function. Also mention the argument definitions. In this example there are two variables, the total and the component, in the percent formula. These are the two arguments for the function.
- The code for the function is simple:

```
Function PerTotal (component, total)
     PerTotal = component/total*100
End Function
```

The first and last line of a function procedure is directly analogous to that of a subroutine. *PerTotal* in this code is the name of the function. *Component* and *total* are the arguments. Pertotal is also the returned value to the spreadsheet. The function contains one line of code, the formula that otherwise would have been entered on the spreadsheet.

- Select the *iserror* sheet and try out the function. Select E4 and choose the Function Wizard tool.
- Select Category **All**, scroll through the Name list box for PerTotal and choose Next. Notice the arguments named in the code are listed in Step 2.
- Tab through the Arguments Input box, selecting cells A2 and D2 for entry, and choose Finish.

This demonstrates how simple it is to write a user-defined function. With the custom function defined, the entry of this calculation anywhere in the spreadsheet has been simplified. User-defined functions are especially advantageous when a complex formula is used regularly in a spreadsheet. Further embellishments of the code are possible, for example, to give the user a Help button and descriptions of the arguments, though these are not the topic of this section. This function is available to workbooks that contain the module sheet with the function code. To be available in any workbook a user-defined function needs to be copied to the *personal.xls*.

Case Study 3.10. Applying a User-Defined Function Globally

- Choose **Window_UnHide** to view the *personal.xls*.
- If UnHide is grayed out, you do not have a *personal.xls* yet. To create a *personal.xls*, record an empty macro and *Store in personal.xls*. Select a worksheet and bring up the Record Macro dialog box. Enter any name and in Options choose *Store in_**Personal Macro Workbook** (Figure 3.2). Do not record any actions; press the Stopbox when it appears. A *personal.xls* should now be created. View it by choosing Window_UnHide.
- With *chap3.xls* and *personal.xls* both open, move to the former workbook and select the macro tab containing the function code. Choose **Edit_Move or Copy** to copy the macro sheet to *personal.xls*.
- Hide *personal.xls* by selecting Window_Hide with this workbook active.
- Close *chap3.xls*, saving your changes.
- Open a new workbook and test out the PerTotal user-defined function. It will be found in the Function Wizard under function category All and *personal.xls!pertotal*.

The creation of user-defined functions had been demonstrated for use in a local workbook or applied globally by placement in the *personal.xls*.

Function procedures are also called by subroutines. An example of using a function in VBA code is when the $\log_{10}$ function is required. The log statement in VBA calculates the natural logarithm of a number. Calculating the $\log_{10}$ requires writing a function procedure. The translation of **answer = $\log_{10}$**(testvalue) to VBA code would be **answer=LOG10(testvalue)** and would need a function procedure for calculation. This function would follow the *Sub . . . End Sub* procedure containing the **LOG10** statement:

```
Function LOG10(X)
      LOG10=LOG(X)/LOG(10#)
End Function
```

When the answer assignment statement is executed, it uses the function procedure to calculate the $\log_{10}$ of testvalue.

3.4. SOLID-STATE CHEMISTRY EXAMPLES USING VBA

3.4.1. Calculation of Bond Distances and Angles

This example is taken from work on the computer simulation of polymer lattices. In solid-state chemistry involving computer modeling of systems it is often necessary to determine the bond lengths, distances, and angles between atom site positions. The simulation program determines relaxed lattice positions in Cartesian coordinates of the atoms in the polymer system. The immediate information ascertained from these results is whether the structure of the polymer has been maintained. Excel affords a quick method for calculating the bond lengths and angles for this determination. The input data for the spreadsheet are the Cartesian coordinates of three consecutive atoms in the polymer chain. We want Excel to calculate the bond lengths or distances of the triad of atoms and the inscribed bond angle. In Cartesian coordinates, the equation for the distance between two points A and B is

$$A–B = \sqrt{(x_a - x_b)^2 + (y_a - y_b)^2 + (z_a - z_b)^2} \tag{3.1}$$

and determination of the inscribed angle C is through rearrangement of the law of cosines:

$$C = \cos^{-1}\left(\frac{c^2 - a^2 - b^2}{-2ab}\right) \tag{3.2}$$

where the symbols have their usual meanings.

- Select the next sheet tab in *chap3.xls* and rename to **Atomic Parameters**.
- Enter an appropriate title for the spreadsheet in row A1, for example, Determination of Atomic Distances and Angles.
- Label a 3 × 3 matrix along the rows as **Atom1, Atom2**, and **Atom3** and along the columns as **X, Y**, and **Z** as demonstrated in the table.

A3 →	X	Y	Z
Atom1			
Atom2			
Atom3			

Start with the left-hand corner cell being A3. This area of the spreadsheet is where the input data is to be entered.

- To make the formulas for calculation easy to follow, assign names to the cells of this matrix. Select B4, the cell representing the x coordinate of Atom1 and choose **Insert_Name_Define**. Type in the name **X_1**. (You cannot use X1 as this would be interpreted as the cell X1 rather than a variable.) This defines X_1 to be the variable name for the B4 cell entry. Continue to select the remaining cells of this matrix, defining the names Y_1, Z_1, X_2, Y_2, Z_2, X_3, Y_3, and Z_3. All the coordinates of Atoms 1, 2, and 3 can now be referred to by their variable names.

The calculations of the bond distances and angle will be placed below this area.

- Enter titles **A1-A2 =, A2-A3 =, A1-A3 =** in cells B8:B10. This is to represent the distances between Atom1 and Atom2, and so on.
- Enter the formulas for the distances between the atoms in C8:C10, using the variable names for coordinates defined above i.e., X_1, Y_1, ... Hence, for the bond length A1–A2 in C8, the formula will be

=SQRT((X_1-X_2)^2+(Y_1-Y_2)^2+ (Z_1-Z_2)^2).

These formulas will result in an answer of 0.0 as there are no values yet entered in the matrix for the coordinates.

- Enter the title **Angle A1-A2-A3** in E9. The formula for the angle will be entered in G9, but first define names for the atomic distances to make this formula easier to read.
- Select cell C8 and define the name **dist12**. Do the same for C9 and C10,

defining the names **dist23** and **dist13**. In this way, these variable names correspond to the *a, b,* and *c* variables in equation (3.2) for calculating the angle.

- Enter the formula for determining the angle described by A1-A2-A3 in G9 with the result being in degrees. The part within the parentheses should be easy enough. To calculate the inverse cosine in Excel use the ACOS function. However, Excel defaults to radians, so the answer must be converted to degrees. This is accomplished by multiplying the formula by **180/PI()**. [The variable for π in Excel is PI().] The complete formula should be

ACOS((dist13^2-dist12^2-dist23^2)/(-2*dist12*dist23))*180/PI().

The calculation entries of the spreadsheet are now finished. To enhance the spreadsheet, further formatting can be performed to highlight the input and output sections. First, alter the numerical display of the data and calculation cells.

- Adjust the decimal display of all input and distance output cells to three decimal places. Do this by highlighting the 3 × 3 matrix of coordinates. Then hold down the CTRL key and highlight the three distance output cells (C8:C10). Select **Format_Cells_Number (tab)**. From this dialog box set Category to **Number** and scroll through the Format Codes to see if the format 0.000 is present. If not, select the 0.00 Format Code and move the mouse cursor into the Code input box and add another zero. In Excel 7, change the Decimal Places list box to **3**. Select OK when choices are complete. This method of adjusting number format is analogous to using the increase–decrease decimal buttons of the formatting toolbar. However, the latter method is reserved for numbers previously entered in the cells.
- Repeat **Format_Cells_Number (tab)** to adjust the angle output cell to one decimal place.

Borders and colors can be added to those cells of interest. This example introduces the various ways the cells can be formatted for augmenting the display of the information in a spreadsheet.

Note: if any of the buttons of the Standard and formatting toolbars are not present in your display of Excel, add them through View_Toolbars_Customize.

- Remove the gridlines of the spreadsheet to make it look less busy. Select **Tools_Options_View (tab)**. Under Window options, clear the **Gridlines** check box. This removes the gridlines from the entire spreadsheet.
- Highlight the 3 × 3 matrix. Press the Borders drop-down button from the formatting toolbar. This displays the Borders toolbar for easy access when formatting borders. There are various buttons for borders in this toolbar. Spend

a little time viewing each type to familiarize yourself with the type of border they place on the cell. From the Borders toolbar, select the outside–inside button by pressing it once. Using the R1C1 format, this button is in the R3C2 position of the Borders toolbar. This convention will be used to help you find the type of border asked for in the following steps.

- Add an outside border to the title of the spreadsheet. There are two choices, a bold and unbold outside border (R3C4, R3C3).
- Add an outside border to the distances titles and output cells by first highlighting these cells (B8:C10). Add bottom borders to cells B8:C8 and B9:C9 (R1C2). Select the distance output cells and align left by pressing the Align Left button from the formatting toolbar. Highlight the range B8:C10 again and select a color for the cell background from the Color button. Likewise, the font color can be changed with the Font Color button.
- Add borders and colors to the angle title and output cells.
- Add a text box to inform the user of the spreadsheet that the coordinates are in angstroms and the angle is in degrees. Click the Text Box button 🖻 of the Drawing toolbar and draw a box on the spreadsheet to desired size. Display the Drawing toolbar by pressing the Drawing tool 🔂 on the Standard toolbar. Enter the text. Don't worry if the text box turns out to be too small to hold the entered text. This can be corrected afterwards. When finished entering text, click outside the text box; pressing ENTER will just keep adding blank lines to the text box. Click the text box to display the handles and adjust the size if necessary. The mouse will become a double arrow when placed over any of the handles. The text box can be moved to any location of the spreadsheet by dragging when the mouse pointer is on any of its sides.

Try out the spreadsheet and enter in data.

- Enter the following atomic coordinates, resulting from a lattice simulation of sodium-doped poly(p-phenylene vinylene), into the spreadsheet:

3.216	0.832	−0.419
3.135	2.106	0.137
2.96	2.258	1.508

- Toggle through the coordinate input matrix with the arrow keys or TAB key and enter the above data.

This spreadsheet operates correctly in that it calculates the bond distances and angles of a triad of atoms. Its operation can be further enhanced by adding VBA code to guide the user inputting the data. The data entry process will be automated to make data entry more efficient and error free. This is done by the InputBox statement in VBA.

Case Study 3.11. Generating a Macro to Prompt the User of a Spreadsheet for Data Entry

- Insert a module sheet and type in the term **InputBox**. Select the term and press F1. This is a quick way to bring up Basic Help. Choose VBA: Input-Box and OK. The syntax for the InputBox statement is described therein. Using Help to look up InputBox will also give the InputBox Method. This is the reference material for the command. The detailed syntax of the commands are left to the reader; only an example of working code is demonstrated. As with all code, it can always be made more efficient. The VBA code in this book may not be the most efficient, but it completes the task at hand. The reader can then go further from here.
- Type in some comment lines at the beginning of the macro sheet describing the purpose of the macro. Don't forget to either delete or comment out the *InputBox* term entered in the above step.
- Begin entering the code for the macro by typing in a **Sub** line, giving the macro a name. The code in Table 3.6 uses the name **InputData**.
- The next few lines will be for:
  - selecting the cell on the spreadsheet for data entry,
  - assigning the output of the input box to a variable, and
  - placing the value of the variable into the selected cell.
- Enter in

```
Range("B4").Select
X1 = InputBox( _
prompt:="Enter x coordinate of Atom1")
ActiveCell.FormulaR1C1 = X1
```

The line continuation syntax in VBA is a space with an underscore. This was used in the inputbox line above. In this code, cell B4 is selected. The InputBox line displays an Input dialog box with the comment in the prompt argument. The user's response is placed in the X1 variable. The last line places the results in the active cell, B4.

- Follow this fragment of code with **End Sub**, select the data sheet, and run the macro to test.

This code now needs to be copied and pasted for the remaining cells of the input section, that is, eight times. In each paste, the range and prompt need to be edited to correspond to the cell at hand. This finished code is shown in Table 3.6. This macro can be run in various ways. So far the Tools_Macro command has been used. As seen in the Record Macro dialog box, it can be assigned to a keypress or a menu item in the Tools menu. It would be easiest for the user to just click a button on the spreadsheet and have the macro run.

Case Study 3.12. Assigning a Macro to a Control Button on the Spreadsheet

- Select the Create Button tool on the Drawing toolbar and describe a small button over an empty area of the spreadsheet.
- After the mouse button is released, an *Assign Macro* dialog box is displayed. Select the name of the macro and choose OK. The control button is still an object that can be edited at this point, as it is framed by size handles. Select the text inside it and change to **Input Data**. If you have already selected outside the control button and it has the appearance of a three-dimensional object, it will need to be selected by holding down the CTRL key or the macro will run.
- When the control button is ready for activating the macro, the mouse pointer will change to a pointing hand when placed over the object. Select the button now and view the running of the macro.
- Add a text box explaining the purpose of the macro button.

The spreadsheet and macro are complete. Figure 3.6 shows the completed spreadsheet. This exercise has shown how the appearance of a spreadsheet can be enhanced and the entry of data aided by a macro.

3.4.2. Calculation of Crystallographic Angles

In this case study, Excel is used to calculate the crystallographic angles of a lattice from the lattice vectors in Cartesian coordinates. This is accomplished through calculation of the directional cosines.

For lattice vectors for the x, y, and z vectors of the unit cell in the form:

X:	X_1	Y_1	Z_1
Y:	X_2	Y_2	Z_2
Z:	X_3	Y_3	Z_3

the formulas are:

$$r_1 = \sqrt{x_1^2 + y_1^2 + z_1^2} \tag{3.3}$$

Table 3.6. Code for Prompting User to Input Data

```
'bond distances and angles macro
'prompts user to enter coordinates into
spreadsheet
'
Sub InputData()

Range("B4").Select
X1 = InputBox( _
     prompt:="Enter x coordinate of
Atom1")
   ActiveCell.FormulaR1C1 = X1
Range("C4").Select
Y1 = InputBox(prompt:="Enter y coordinate
of Atom1")
   ActiveCell.FormulaR1C1 = Y1
Range("D4").Select
Z1 = InputBox(prompt:="Enter z coordinate
of Atom1")
   ActiveCell.FormulaR1C1 = Z1
Range("B5").Select
X2 = InputBox(prompt:="Enter x coordinate
of Atom2")
ActiveCell.FormulaR1C1 = X2
Range("C5").Select
Y2 = InputBox(prompt:="Enter y coordinate
of Atom2")
ActiveCell.FormulaR1C1 = Y2
Range("D5").Select
Z2 = InputBox(prompt:="Enter z coordinate
of Atom2")
ActiveCell.FormulaR1C1 = Z2
Range("B6").Select
X3 = InputBox(prompt:="Enter x coordinate
of Atom3")
ActiveCell.FormulaR1C1 = X3
Range("C6").Select
Y3 = InputBox(prompt:="Enter y coordinate
of Atom3")
ActiveCell.FormulaR1C1 = Y3
Range("D6").Select
Z3 = InputBox(prompt:="Enter z coordinate
of Atom3")
ActiveCell.FormulaR1C1 = Z3

End Sub
```

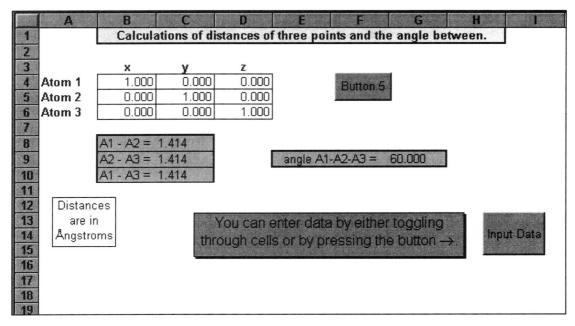

Figure 3.6. The spreadsheet for calculating bond distances and angles.

$$l_1 = \frac{x_1}{r_1}$$

$$m_1 = \frac{y_1}{r_1}$$

$$n_1 = \frac{z_1}{r_1}.$$

The formulas for r_2, r_3, l_2, l_3, m_2, m_3, n_2, n_3 are calculated similarly with the variables of the corresponding subscript. The variables l, m, and n are known as the direction cosines and are from which the (crystallographic) angles are calculated:

$$\cos \Theta_{23} = l_2 l_3 + m_2 m_3 + n_2 n_3$$
$$\cos \Theta_{13} = l_1 l_3 + m_1 m_3 + n_1 n_3 \qquad (3.4)$$
$$\cos \Theta_{12} = l_1 l_2 + m_1 m_2 + n_1 n_2$$

These are also known as angles α, β, and γ. The lattice vectors in Cartesian format is the data to be entered into the spreadsheet, and the output is the angles by way of the directional cosines.

- Select the next sheet and give a descriptive name.
- Enter an appropriate title to the spreadsheet in row A, for example, **Calculation of Crystallographic Angles**.
- Label a 3 × 3 matrix along the rows as **X Vector, Y Vector, Z Vector** and along the columns as **X, Y, Z** as demonstrated in the Table.

B3 →	X	Y	Z
X Vector			
Y Vector			
Z Vector			

This entails highlighting a 4 × 4 matrix. Start with the left-hand corner cell being B3. This area of the spreadsheet is where the input data is to be entered.

The steps for calculation of the directional cosines follow.

- Enter r_1 = in cell A8. Do the same for r_2 and r_3 in cell A9:A10. Highlight the range A8:A10 and **Align right** with its button in the formatting toolbar.
- Place the formulas for these variables in B8:B10. Select B8 and enter the formula for r_1. Select the cell addresses of the variables in the formula. Hence, the entry for B8 should be **=SQRT(C4^2+D4^2+E4^2)**.
- Highlight B8:B10 and select **Edit_Fill_Down** (CTRL + D). Examine the contents of these cells to ensure that the formulas are correct. Reselect this range and **Align left**.
- Enter the titles for the *l, m,* and *n* subscripted variables in the ranges C8:C10, A12:A14, and C12:C14, respectively, in the same manner as for the *r* variables. Do not forget to **Align right**.
- Place the formulas for the *l* variables in D8:D10. In D8 enter the formula for l_1. Select the cell addresses of the variables in the formula. This is **=C4/B8**. Highlight D8:D10 and **Fill_Down**. Reselect this range and **Align left**.
- Place the formulas for the *m* variables in B12:B14 in the same manner as for the l variables.
- Place the formulas for the *n* variables in D12:D14 in the same manner described above.

The output section of the spreadsheet is a 3×2 matrix for the angle labels and their values.

- Type in **Alpha, Beta, Gamma** in G4:G6. Format **Bold** and **Align right**.
- Enter the formulas for the angles in H4:H6. These cells need to be entered individually; their entry cannot be aided with Fill_Down as can be seen by the nature of the variable listings in the formulas. [*Hint:* H4 entry is **=ACOS(D9*D10+B13*B14_D13*D14)*(180/PI).**]

Alter the numerical display of the data and calculation cells as follows.

- Adjust the number of decimal places of the lattice vector input cells and the directional cosine calculation cells to four places. Highlight all these cells, holding down the CTRL key to select nonadjacent cells as necessary. Select **Format_Cells_Number (tab)** or press CTRL + 1. From this dialog box choose **Number** from the Category list and scroll through the Format Codes to see if the format 0.0000 is present. If not, select the 0.00 Format Code and move the mouse cursor into the Code input box and add two more zeros. Select OK when choices are complete.
- Repeat **Format_Cells_Number (tab)** to adjust the angle output cells to one decimal place.

Steps for enhancing the display of the spreadsheet by adding borders and colors to those cells of interest follow.

- Remove the gridlines of the spreadsheet from Tools_Options_View (tab). Under Window options clear the Gridlines check box. This removes the gridlines from the entire spreadsheet.
- Select the Cartesian vectors matrix and place a border inside and outside.
- Highlight the angles title and output cells and place a border inside and outside. Also with this selection, change the background color to maroon and the font color to yellow with their buttons from the formatting toolbar.
- Place a border around the title cells. Enhance the title by changing the background color and the font color if so desired.
- Hide the cells that contain calculation of the direction cosines. Highlight these cells and change the font color to white. The spreadsheet displays only those cells of interest.

Figure 3.7 shows the formatting results of the spreadsheet.
The following table contains lattice vectors for the results of a lattice simula-

tion of poly(*p*-phenylene vinylene). Enter in the lattice vectors and then, calculate the crystallographic angles using the constructed spreadsheet.

X:	8.779	0.045	0.019
Y:	−0.023	6.519	−4.752
Z:	−0.002	−2.179	6.262

- Toggle through the lattice vectors input matrix with the arrow keys or TAB key and enter the above data.

The data entry into this spreadsheet can be aided with an input box macro similar to that used in the previous case study. Insert a module sheet and build a macro to prompt the user through inputting the data for each of the lattice components. Compare your macro with that in Table 3.7. Select the data sheet and draw a control button on the spreadsheet and assign the macro to it. Try out the macro by pressing the control button and stepping through the input boxes.

These two case studies show how a spreadsheet with involved formulas of a multistep process can be formatted to display only the input and output data cells, effectively hiding the formula cells from the user. Furthermore, the data entry is aided by macros to ensure correct data entry. The combination of spreadsheet formatting and simple VBA code enhances appearance and usage, resulting in a user-friendly spreadsheet.

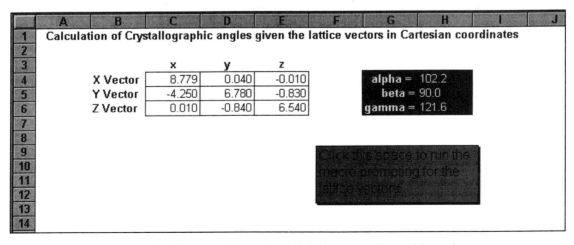

Figure 3.7. The spreadsheet for calculating crystallographic angles.

Table 3.7. Code for Prompting User through Data Entry for Calculation of Crystallographic Angles

```
'crystallographic angles macro
'this macro asks the user to input the lattice vectors
'to calculate the crystallographic angles

Sub XtalAngles()
Range("c4").Select
X1 = InputBox(prompt:="Enter x component of x
lattice vector")
ActiveCell.FormulaR1C1 = X1
Range("d4").Select
Y1 = InputBox(prompt:="Enter y component of x
lattice vector")
ActiveCell.FormulaR1C1 = Y1
Range("e4").Select
Z1 = InputBox(prompt:="Enter z component of x
lattice vector")
ActiveCell.FormulaR1C1 = Z1
Range("c5").Select
X2 = InputBox(prompt:="Enter x component of y
lattice vector")
ActiveCell.FormulaR1C1 = X2
Range("d5").Select
Y2 = InputBox(prompt:="Enter y component of y
lattice vector")
ActiveCell.FormulaR1C1 = Y2
Range("e5").Select
Z2 = InputBox(prompt:="Enter z component of y
lattice vector")
ActiveCell.FormulaR1C1 = Z2
Range("c6").Select
X3 = InputBox(prompt:="Enter x component of z
lattice vector")
ActiveCell.FormulaR1C1 = X3
Range("d6").Select
Y3 = InputBox(prompt:="Enter y component of z
lattice vector")
ActiveCell.FormulaR1C1 = Y3
Range("e6").Select
Z3 = InputBox(prompt:="Enter z component of z
lattice vector")
ActiveCell.FormulaR1C1 = Z3
End Sub
```

This chapter has introduced the basic concepts of VBA through examples. VBA programming was introduced in a "hands-on" fashion to bring the reader quickly into a programming frame of mind. This approach was intentional, since it demonstrates to the user how quickly the basics of VBA can be picked up. You will learn further VBA techniques as you go through later case studies. In the *chap3.xls* file there is a sheet named *Calib Data* and an accompanying macro. Explore these two sheets to practice your skills and determine how they operate. The reader is advised to consult the *Microsoft Excel User's Guide* and *Microsoft Excel Visual Basic User's Guide* for further information on programming in VBA.

CHAPTER 4

CASE STUDIES IN QUANTUM CHEMISTRY

Excel is an excellent vehicle for computing complex formulas. The field of quantum chemistry with its Schrödinger wave equation provides numerous examples of such formulas. The following case studies demonstrate how Excel easily copes with their complexity, at the same time allowing the user a means to better understand the underlying theory through visualization of the results.

4.1. QUANTUM LEAKS

Any student of quantum chemistry is well familiar with the discussions of the particle in a box with walls of infinite potential. The concept of quantum leaks stems from the scenario that if the potential of the walls is not of infinite potential then a striking particle may penetrate it. Furthermore, if the walls are of limited width, then the amplitude of the particle's wavefunction might not fall to zero before it reaches the other side of the wall where the potential is low. Quantum leaks are an example of a classical forbidden zone, a particle being outside the box even though classical physics predicts it has insufficient energy to escape, and occurrence favors low mass projectiles. An aspect of this discussion is that there exists a probability that a particle will either penetrate or be reflected from a less than infinite potential barrier and is dependent on the mass of the particle. This case study examines the probability that a particle will be reflected back from a dip in an otherwise uniform potential well.

Case Study 4.1: Plotting the Probability Function for a Particle

Plot the probability function for a particle reflected back from a dip in an otherwise uniform potential as a function of λ. Take the depth of the well as $V = 5$ eV and its width as $L = 0.1$ nm. Plot the probability as a function of E/V for a proton and deuteron [1, 2].

The probability of reflection is equal to (1 – probability of penetration) and is expressed as

$$P(\text{reflection}) = \frac{G}{(1 + G)} \tag{4.1}$$

where

$$G = (E/V)^2 \sin^2 \frac{\left\{\left[\dfrac{2m(E + V)}{\eta^2}\right]^{\frac{1}{2}} L\right\}}{4[1 + (E/V)]} \tag{4.2}$$

Substituting in λ for E/V reduces the formula for G to

$$G = \sin^2 \frac{\left\{\left[\dfrac{2mV}{\eta^2}(1 + \lambda)\right]^{\frac{1}{2}} L\right\}}{4\lambda(1 + \lambda)} \tag{4.3}$$

Excel is used to calculate the value of G as a function of λ. To simplify the formula entry, the term $(2mV/\eta^2)^{\frac{1}{2}} \cdot L$ is calculated first and is denoted A in the formula below. $G(\lambda)$ becomes:

$$= \frac{\sin^2\{A(1 + \lambda)^{\frac{1}{2}} L\}}{4\lambda(1 + \lambda)} \tag{4.4}$$

With the mathematical preliminaries finished, the spreadsheet is ready for construction.

- Open a new workbook and Save As *Chap4.xls*.
- Rename Sheet1 to *probability*.
- Enter a title in A1.
- Calculate the value of the A parameter as $(2mV/\eta^2)^{\frac{1}{2}} \cdot L$ for the proton and deuteron in cells A3:A4. Convert the V and L variables to SI units: joules and meters, respectively. Select A3 and A4 in turn and define the variable names **Ap** and **Ad,** respectively, to them.

- In A6:E6 enter a title row for **λ, G-proton, G-deuteron, P-proton, P-deuteron.**
- Use Edit_Fill_Series to enter values for λ from 0.01 to 3, incrementing 0.01, starting in cell A7.
- In B7 enter the formula for G for the proton according to equation (4.4), using its defined name for the *A* parameter. Split the window and select the entire intended range for column B, that is, what corresponds to data in column A. This is demonstrated in Figure 4.1.
- Copy the formula in B7 to C7. Adjust the *A* parameter in the formula to denote the deuteron. The copy action has changed the cell references for λ to column B; make the appropriate edits to represent column A.
- Fill_Down the formula for the deuteron in column C.
- Enter the formula for the probability, P(G), in D7 and fill the formula down through the appropriate range.
- Enter the formula again in E7 (or copy from D7) and Fill_Down as above.

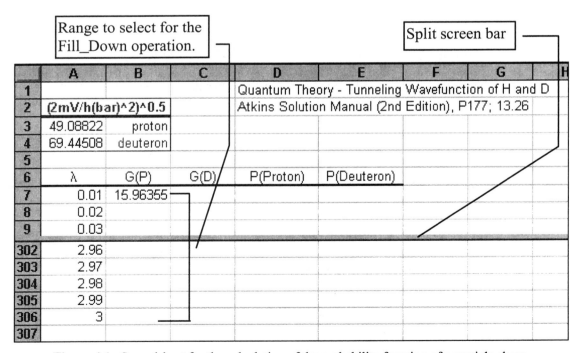

Figure 4.1. Spreadsheet for the calculation of the probability function of a particle showing the intended range for the formula in the Fill_Down operation.

The probability as a function of E/V is constructed from the completed spreadsheet. The formula entries should very similar to that shown in Figure 4.2.

- Select columns A, D, and E for graphing by splitting the screen and scrolling the bottom pane to reveal the end of the data entries. Select A6, hold down the SHIFT key, select A306, then release the SHIFT key. Hold the CTRL key down and select D6. Release the CTRL key, hold down the SHIFT key, and select E306.
- Press the Chart Wizard tool and draw the outline of the graph.
- Choose an XY Scatter type graph (Step 2) with a grid background (Step 3). Ensure data is in columns, use first row and column, and label the graph appropriately in the last two steps. Choose Finish.
- Further edits can be done to enhance the look of the graph. Activate the graph and select the plot area. Press CTRL + 1 or choose **Format_Selected Plot Area.** With any object or attribute of the graph selected, using the CTRL + 1 keypress brings up the related formatting dialog box. From this dialog box select **None** for the plot area and choose OK. Select one of the grid lines and press CTRL + 1. In the Patterns tab, change the color to gray and choose OK. Select an opposite grid line and change its color to gray. Format the axes to ensure the x and y scales range from 0 to 3 and 0 to 1, respectively. Select one of the plotted data series and edit the Patterns tab to reflect no markers, change the line to a heavier weight, and check the Smoothed Line box. Do the same for the other data series. Change their colors in this dialog box if desired. Drag the legend box inside the plot area and extend the plot area to the width of the marquee.

The graph is complete and should look similar to that shown in Figure 4.3. This plot shows the oscillating decaying wavefunction of the proton and deuteron as a function of wavelength. The proton curve is oscillating with a

5				
6	λ	G(P)	P(Proton)	P(Deuteron)
7	0.01	=(SIN(Ap*(1+A7)^0.5))^2/(4*A7*(1+A7))	=B7/(1+B7)	=C7/(1+C7)
8	0.02	=(SIN(Ap*(1+A8)^0.5))^2/(4*A8*(1+A8))	=B8/(1+B8)	=C8/(1+C8)
9	0.03	=(SIN(Ap*(1+A9)^0.5))^2/(4*A9*(1+A9))	=B9/(1+B9)	=C9/(1+C9)
10	0.04	=(SIN(Ap*(1+A10)^0.5))^2/(4*A10*(1+A10))	=B10/(1+B10)	=C10/(1+C10)

Figure 4.2. Partial spreadsheet for the calculation of probability of reflection as a function of wavelength.

greater frequency than the deuteron, as expected due to their differences in mass. The use of the Excel spreadsheet has simplified the process of manipulating and graphing the probability data.

4.2. PLOT OF THE RADIAL DISTRIBUTION FUNCTION

The Schrödinger wave equation is solved for a quantum number set configuration yielding their related energies and wavefunctions. The wavefunction for an electron in an atom describes an atomic orbital, and each unique set of quantum numbers corresponds to a set of orbitals. Much information is derived from the interpretation and application of wavefunctions. A physical interpretation of the wavefunction is determining the most probable distance of an electron from the nucleus. This is achieved through the radial distribution function (RDF), defined as the probability of finding an electron at a radius r from the nucleus:

$$RDF = 4\pi^2 r^2 \psi^2 \tag{4.5}$$

A variable exercise in the study of quantum chemistry is to plot the RDF for an electron in a particular orbital vs. the radius of the electron from the nucleus of the atom. The maximum value of the RDF therefore defines the most probable radius of the electron lying in that orbital. This radius can be determined from inspection of the plot and/or the first derivative of the RDF function.

Case Study 4.2. Determining the Most Probable Distance of an Electron

Determine the most probable distance of an electron in the 3 s orbital of hydrogen, from the nucleus, [1, 2]. The radial dependence, R_{nl}, of the 3 s orbital is (from Atkins [1]),

$$R_{30}(r) = \left(\frac{Z}{a_o}\right)^{3/2} \left(\frac{1}{9\sqrt{3}}\right)(6 - 6\rho + \rho^2)e^{(-\rho/2)} \tag{4.6}$$

where $a_o = 52.92$ pm, $\rho = 2Zr/3a_o$, and Z is the atomic number. For the purpose of calculating the most probable radius, the RDF can be approximated as

$$RDF \propto 4\pi r r^2 R_{30}^2(r) \propto r^2 (6 - 6\rho + \rho^2)^2 e^{-\rho} \tag{4.7}$$

$$\propto \rho^2(6 - 6\rho + \rho^2)^2 e^{-\rho} \tag{4.8}$$

This simplifies the plot to RDF vs. ρ.

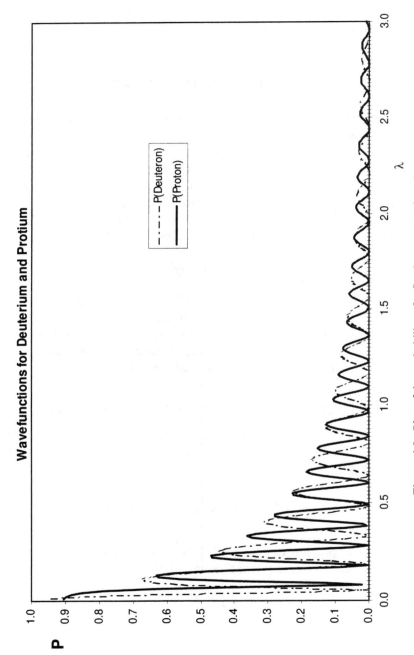

Figure 4.3. Plot of the probability of reflection *vs.* wavelength.

- Select the next available sheet tab and rename to *RDF.*
- Enter a title in row A and other details of the problem.
- In cell A6:B6, enter in data titles ρ and **RDF.**
- Enter values for ρ ranging from 0 to 15, incrementing every 0.1 through Edit_Fill_Series.
- Enter the formula for RDF according to equation (4.8). Figure 4.4 shows this formula entry.
- Fill the formula down the column to correspond to the ρ entries of column A.
- To mark the maximum in the RDF function more clearly, enter another column title for the first derivative, **d(RDF)/dρ,** in E6.
- In E7 enter the formula **=(B7–B6)/0.1.** Fill this formula down to the end of entered data.

Graphing

- Highlight the ρ and RDF data and select the Chart Wizard tool to draw an XY Scatter graph of format type 2.
- Ensure the entries in Step 4 are correct and label the graph in Step 5.
- Format the x axis by double clicking it. Select the Scale tab and adjust the scale so the major unit is **1** and the minor tick mark is **0.2.** Select the Patterns tab and in the Tick Mark Type group select the **outside** radio button for the minor tick marks.
- Select the data series and change the color and mark if the Excel defaults are not to your preference. Change the Background of the Marker to **none.**
- Select the **d(RDF)/dρ** data on the spreadsheet; be sure to include the title row. The entire range is C6:C156. Copy the data. Select the graph and choose **Edit_Paste.** The new data should have been automatically entered

5	ρ	RDF
6	0	=A6^2*(6-6*A6+A6^2)^2*EXP(-A6)
7	0.1	=A7^2*(6-6*A7+A7^2)^2*EXP(-A7)
8	0.2	=A8^2*(6-6*A8+A8^2)^2*EXP(-A8)
9	0.3	=A9^2*(6-6*A9+A9^2)^2*EXP(-A9)

(a)

5	ρ	RDF
6	0	0
7	0.1	0.26482872
8	0.2	0.767170365
9	0.3	1.227068335

(b)

Figure 4.4. Partial spreadsheet displaying the RDF formula entry. (a) Formula entries for the RDF function for the ρ values. (b) The resulting values.

into the graph. If the new data series is of different range or from a different area of the spreadsheet a Paste Special dialog box for graphing may
appear. This was shown in Figure 1.27.

- With two different functions or plot series on the same graph, it is often
best to have a second y axis or double y plot. Activate the graph and format the new data series [d(RDF)/dρ] by double clicking its plot line or using the CTRL + 1 keypress. Select the Axis tab and choose Plot Series On
Secondary Axis.

The final graph should be similar to that shown in Figure 4.5. This graph
can be further enhanced with Drawing tools: adding a line to mark zero on the
second y axis and an arrow to highlight the ρ value corresponding to the most
probable distance of the 3 s electron.

- Activate the graph and display the Drawing tool bar by either selecting its
button on the Standard toolbar or through View_Toolbars. Select the line
button and place the crosshairs over the zero of the second y axis. Hold
down the SHIFT key while clicking the line tool and dragging the line the
width of the plot area. The purpose of holding down the shift key is to ensure a straight line. Compare the difference of drawing a line with and

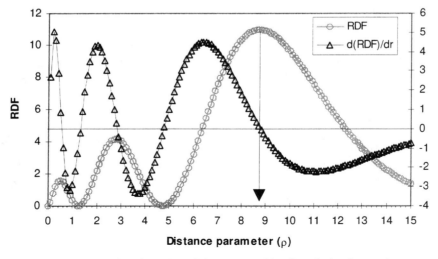

Figure 4.5. The double y plot of the RDF and its first derivative against ρ.

without the SHIFT key held down. Select the arrow button and draw a line in the same manner, highlighting the maximum in the RDF function.

The last step in this task is to evaluate r for which the RDF is a maximum. The graph demonstrates that this is so at $\rho = 2Zr/3a_o = 8.8$. Use the spreadsheet to determine r; enter the formula into a cell. Your answer should be 698 pm.

4.3. MOLECULAR STRUCTURE

Molecular orbitals are linear combinations of atomic orbitals rendering the same number of orbitals that were combined. Molecular orbital is addition of ψ_+ and ψ_-, repulsive and attractive components. Molecule A-B will have

$$\text{Bonding orbital: } \psi_+ = 1s_A + 1s_B \tag{4.9}$$

$$\text{Antibonding orbital: } \psi_- = 1s_A - 1s_B \tag{4.10}$$

with

$$1s_A = 1s_B \propto e^{(-r/a_o)} \tag{4.11}$$

where the distance r is measured from atom A. The same formula applies to $1s_B$ and here r is measured from atom B. To have a consistent distance variable, z is defined as the distance measured from A toward B for both atoms:

$$\psi_\pm \propto e^{(-|z|/a_0)} \pm e^{-|z-R|/a_0} \tag{4.12}$$

with the distance, z, measured from atom A.

4.3.1. Molecular Orbitals

For complete understanding of molecular orbital theory, graphic examples are necessary to visualize the involved formulas required. This next task demonstrates the application of molecular orbitals.

Case Study 4.3. Plotting the Amplitude of Bonding and Antibonding Molecular Orbitals Along the Internuclear Axis

For the molecule A-B with bond length 106 pm, the steps are as follows.

- Select the next sheet tab and rename to *orbitals*.

- Enter titles in the first few rows.
- Define names for the Bohr radius and bond length. In F2:F3 enter the text **A$_0$** and **R**. In cells G2:G3 enter their values, 52.9 and 106 pm, respectively. Do not enter their units here; place them in H2:H3. Select each numerical entry in turn and choose **Insert_Name_Define** to register these values with their names. Their variable names will then be used in the formulas for the wavefunctions.
- In A5:E5 enter the data column titles: **z (pm), |z| (pm), |z-R|, ψ_+, ψ_-.**
- Enter data ranging from −100 to 200 for the z values by using Edit_Fill_Series in column A.
- Use the absolute (ABS) function in column B to calculate the absolute value of the z entries of column A. Select cell B6, choose the Function Wizard and select the **ABS** function. Select A6 for the entry in Step 2. Fill_Down the entries of column B.
- In C6 enter the formula for $|z–R|$ and Fill_Down.
- The formula for the bonding goes in D6 according to equation 4.9. Select D6 to enter the bonding calculation and instead of using the Function Wizard, type in =**EXP (** and then the keypress CTRL + A. This brings up the Function Wizard in Step 2 for the EXP function. Type in the formula -**B6/A$_0$** and press Finish. This brings the cursor back to the formula bar. Continue the formula by entering +**EXP(-C6/A$_0$)** and press RETURN. Fill_Down the formula.
- Enter the antibonding wavefunctions in E6 according to equation 4.10 and Fill_Down.

A sample of the formulas and the results are shown in Figure 4.6 for the bonding wavefunction.

Graphing

- Select the data in the **z, ψ_+,** and **ψ_-** columns and press F11. This will take you through the Chart Wizard and, on finishing, it will place the graph on a separate sheet, a chart sheet. Choose an XY Scatter type of format 2 and label the graph in the Chart Wizard steps.
- Format any of the chart attributes to accent the plot. Add arrows to label the node and antinodes. Compare your graph to Figure 4.7. Atom A is at z = 0 and Atom B is at z = 106 pm and the graph shows the ψ_+ at a maximum at these two points. The antibonding orbital wavefunction has a node midway between the atoms corresponding to a minimum in the bonding orbital wavefunction.

5	z(pm)	|z|	|z-R|	ψ+
6	-100	=ABS(A6)	=ABS(A6-Ra)	=EXP(-B6/Ao)+EXP(-C6/Ao)
7	-95	=ABS(A7)	=ABS(A7-Ra)	=EXP(-B7/Ao)+EXP(-C7/Ao)
8	-90	=ABS(A8)	=ABS(A8-Ra)	=EXP(-B8/Ao)+EXP(-C8/Ao)

(a)

5	z(pm)	|z|	|z-R|	ψ+
6	-100	100	206	0.171378438
7	-95	95	201	0.188366994
8	-90	90	196	0.207039608

(b)

Figure 4.6. Partial spreadsheet of the formulas of the molecular bonding orbital. (a) Formula entries. (b) Corresponding formula results.

4.3.2. Electron Densities

Electron density plots of the molecular orbitals display the probable distribution of the bonding electrons between the joined atoms. The formulas for the electron density $\rho_\pm$ for bonding and antibonding orbitals [1, 2] are given as

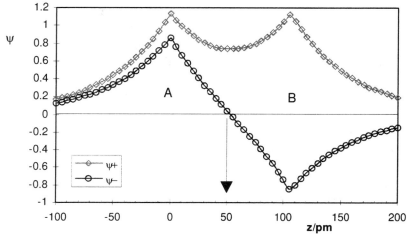

Figure 4.7. Orbital plot.

$$\rho_{\pm} = N^2 \, (1s_A \pm 1s_B)^2 \tag{4.13}$$

An expression for the sum of the electron densities ρ is

$$\rho = N^2 \, (1s_A^2 + 1s_B^2) = N^2 \, [\exp(-2|z|/a_o) + \exp(-2|z - R|/a_o)] \tag{4.14}$$

N is the normalizing factor in both equations 4.1 and 4.2.

Case Study 4.4. Plotting the Electron Density Along the Internuclear Axis

Use the orbitals constructed in the above task. Normalize ρ_+ with 1218 $pm^{3/2}$ and ρ_- with 622 $pm^{3/2}$. Also plot the sum of the electron densities ρ and here take N^2 to be $1/(9.35 \times 10^{-5}) \, pm^3$ [1].

- Select the next sheet tab and rename to *electron density*.
- Enter descriptive information on the task at hand in the first two rows.
- Make N^2 a named variable through Insert_Name_Define. In F2 place N^2 and its value in G2. Select G2 and define as a named variable. Use the name **Nsqrd**. N2 cannot be used as a name because it is also a valid Excel cell address.
- Starting with A5, enter across the data column headings of **z, ρ_+, ρ_-, ρ, $\delta\rho_+$, $\delta\rho_-$.**
- Enter data for z ranging from -100 to 200, incrementing by 5, through the Edit_Fill_Series menu. Alternatively, this range can be copied and pasted from the Orbital spreadsheet.
- The formulas for the ρ parameters will be entered by selecting the ψ function cells of the Orbital sheet, not by reentering the original data. This is a powerful aspect of a spreadsheet. Start with ρ_+ according to equation (4.13). Select B6, then the equal sign, and open parenthesis =(. Select the Orbital sheet and then D6. This brings the ψ_+ value into this formula. Finish the rest of the formula with the closed parenthesis, squaring this, and dividing by the normalizing value of 1218 $pm^{3/2}$ squared. Fill_Down the formula.
- Repeat this formula for ρ_- in C6. Normalize with 622 $pm^{3/2}$ by dividing by this number squared.
- Use equation (4.14) to enter the formula for ρ in E6. Use the named variable for N^2 and select the Orbital sheet for the $|z|$ and $|z-R|$ values. A partial spreadsheet of these formula entries with all their results can be seen in Figure 4.8.

Graphing

- Select the z, ρ_+, ρ_-, and ρ data for graphing. Split the screen and select A5.

z/pm	ρ_+ (pm^{-3})	ρ_- (pm^{-3})	ρ (pm^{-3})
-100	=(Orbitals!D6)^2/1218^2	=(Orbitals!E6)^2/622^2	=Nsqrd*(EXP(-2*Orbitals!B6/Ao)+EXP(-2*Orbitals!C6/Ao))
-95	=(Orbitals!D7)^2/1218^2	=(Orbitals!E7)^2/622^2	=Nsqrd*(EXP(-2*Orbitals!B7/Ao)+EXP(-2*Orbitals!C7/Ao))
-90	=(Orbitals!D8)^2/1218^2	=(Orbitals!E8)^2/622^2	=Nsqrd*(EXP(-2*Orbitals!B8/Ao)+EXP(-2*Orbitals!C8/Ao))
-85	=(Orbitals!D9)^2/1218^2	=(Orbitals!E9)^2/622^2	=Nsqrd*(EXP(-2*Orbitals!B9/Ao)+EXP(-2*Orbitals!C9/Ao))

(a)

z/pm	ρ_+ (pm^{-3})	ρ_- (pm^{-3})	ρ (pm^{-3})
-100	1.9798E-08	4.412E-08	2.48352E-08
-95	2.3917E-08	5.331E-08	3.0003E-08
-90	2.8894E-08	6.44E-08	3.62461E-08
-85	3.4907E-08	7.78E-08	4.37884E-08

(b)

Figure 4.8. Partial spreadsheet showing the formula entries for the electron density calculations.

Hold down the SHIFT key while selecting D66. Select the Chart Wizard tool.

- Choose an XY Scatter graph with format 2. Label the graph accordingly.
- Double click the x axis and select the Scale tab. Change the Value (Y) Axis Crosses at to **-100.** This moves the y axis to the right edge of the plot area.
- Bring up the same formatting dialog box for the y axis. Change the Minimum value and Value (X) Axis Crosses at to **-2E-7.** Select the Number tab and choose **Scientific.** These edits augment the appearance of the y axis with respect to the plotted data.
- Double click the Legend box and in the Font tax change the font to Symbol in order for the ρ symbol to be displayed in the legend.
- Format the data series markers to your preference in style and color. Compare your plot of the electron density with Figure 4.9.

The next task is a continuing exercise on electron density in order to identify the shifts of electron density that occur in bond formation. This is constructing the difference density plot $\delta\rho_+$ and $\delta\rho_-$ as a function of internuclear distance. The formulas for $\delta\rho_+$ and $\delta\rho_-$ are as follows:

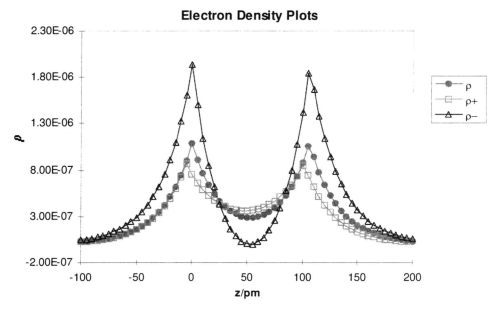

Figure 4.9. Electron density plot.

$$\delta\rho_+ = \rho_+ - \rho \qquad (4.15)$$

$$\delta\rho_- = \rho_- - \rho \qquad (4.16)$$

Case Study 4.5. Plotting Change in Electron Density

Plot the change in electron density that occurs in the formation of a bond and an antibond by subtracting the relevant curves.

- Select the *electron density* sheet tab.
- Enter in data column headings for $\delta\rho_+$, $\delta\rho_-$ adjacent to the existing electron density headings (E5:F5).
- In the cell under the title cell, enter the formula for each by selecting the appropriate previous cells according to the formulas in equations (4.15) and (4.16). Fill_Down the columns.

Graphing

- Select the **z**, $\delta\rho_+$, $\delta\rho_-$ data and choose the Chart Wizard tool. After selecting the *z* data, remember to hold the CTRL key down to select the nonadjacent $\delta\rho_+$ and $\delta\rho_-$ data as outlined in previous tasks.

- Choose an XY Scatter graph of format 2. Ensure the correct selections have been made in Step 4 and give the graph a title and an x-axis label in the final step.

- Activate the graph and double click the x-axis. In the Format Axis dialog box, select the **Scale** tab and Value (Y) Crosses at **-100.** Enter **25** in the Minor Unit input box. Select the **Patterns** tab and choose the Minor Tick Mark type to Outside. The color and font can also be changed in this dialog box.

- Double click the y axis. Again select the **Scale** tab and Value (X) Crosses at **-4E-7,** the same value Excel has defaulted to for the Minimum. Select the **Number** tab and choose **Scientific** and one decimal place.

- Double click the plot area with the pointer on the top border. In the Format Chart Area Dialog box choose **None** for Border.

- Double click the Legend box and in the Font tab change the font to Symbol in order for the Greek symbols to be displayed in the legend. Select the Legend box and move to a more suitable location if necessary.

Compare your graph with that shown in Figure 4.10.

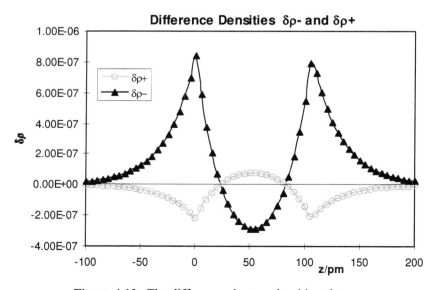

Figure 4.10. The difference electron densities plots.

These case studies and their graphs have shown how Excel greatly simplifies the computations and graphing of complex formulas. This is by no means the limit of Excel capabilities. The extent of the cell entry for a formula is vast and what cannot be achieved through the formula bar can be through VBA. With the basic techniques for formulas and computations in Excel, you should be able to carry this knowledge further to your own needs.

REFERENCES

[1]   Atkins, P. W., *Physical Chemistry,* 2nd ed. Oxford University Press, 1982.

[2]   Atkins, P. W., *Solutions Manual for Physical Chemistry,* 2nd ed. Oxford University Press, 1982.

CHAPTER 5

CASE STUDIES IN
PHYSICAL CHEMISTRY

5.1. INTRODUCTION

As a subject area, physical chemistry is primarily about relating the observed chemistry of a system under study to physical laws. It is mathematically intensive and can prove a difficult subject for many students. Spreadsheets can play an important role in bringing the mathematics to life, enabling students to play interactively with equations and simultaneously observe the graphical consequences of changing variables within the equations. There are many excellent general texts on physical chemistry, such as [1], that can provide numerous examples for spreadsheet studies, and other authors have covered using spreadsheets for physical chemistry calculations in considerable depth [2, 3], so in this chapter we will focus on how to use Excel V.7.0 to investigate specific topics and we will provide enough background to enable users to substitute their own problems.

We have already shown how Excel can be used to graph quite complex mathematical formalisms with examples taken from quantum chemistry. In this chapter, we will take some common topics in physical chemistry and show how a spreadsheet approach can enable students to explore quickly the effect of variation of equation parameters on the graphical display and how this can be interpreted in terms of the underlying chemistry.

5.2. ACTIVITY AND CONCENTRATION

5.2.1. Background

Activity may be regarded as "effective concentration," a concept that arises from attempting to define the impact of surrounding ions on a particular ion's effect on its environment. Two extremes can be identified.

1. Isolated ions (i.e., at infinite dilution) that can exert their full influence in terms of electrostatics.
2. Dense ion population (i.e., concentrated electrolyte) in which each individual ion's effect is shielded to some extent from the total environment by ions of opposite charge, which tend to congregate around a counter ion.

Hence one can predict that activity will tend to decrease with increasing ionic strength. The relationship between ionic strength and single ion activity can be explored by means of the Davies equation (an extended form of the Debye–Huckel equation):

$$-\log f_i = A z_i^2 \left[\frac{\sqrt{I}}{1 + \sqrt{I}} - 0.2\, I \right] \tag{5.1}$$

where z_i = the charge on an ion, i,
   A = a constant (0.512 in the case of water),
   f_i = the activity coefficient of i,
   I = the ionic strength, which is defined as

$$I = 0.5 \sum c_i z_i^2 \tag{5.2}$$

where c_i is the concentration of any ion i, and z_i is its charge. The activity coefficient relates activity and concentration via the equation

$$a_i = f_i c_i \tag{5.3}$$

from which it is clear that as long as $f_i \Rightarrow 1$, concentration and activity will be equal.

Case Study 5.1 Preparing a Worksheet to Investigate the Effect of Electrolyte Type and Electrolyte Concentration on the Activity Coefficient of the Cation in a Solution

Suggested Method

- Open a new workbook, and begin by defining named variables that will be used in the calculations (e.g., **Za, Zb,** and **MAGZb** for the charge on the cation **a,** the anion **b,** and the magnitude of the charge on the anion, respectively). Figure 5.1 shows a layout that has been used in the workbook *activity.*

- Enter the range of electrolyte concentration (as log C) over which the calculations are to be made (e.g., 10^{-6} M to 1.0 M)—this can be easily achieved by entering **–6** in the first cell of the range (e.g., B7), and then using the Fill_Down command with the increment set to **0.25** and the final value to **1.0.** A logarithmic scale is recommended because of the wide range involved.

- Now calculate the concentrations equivalent to these logarithms using the formula =**10^(first cell),** where *first cell* is the address of the cell containing the first number in the logarithmic range. Enter the formula in the adjacent cell in the next row and fill the formula down over the range of concentrations (see Figure 5.1).

- Calculate the ionic strength in the next column using equation (5.2) by entering a formula such as =**0.5*((A7*MAGZb*Za^2)+(A7*Za*Zb^2))**

	A	B	C	D	E	F
1	Named Variables					
2	Za	2				
3	Zb	-1				
4	MAGZb	1				
5						
6	Cab	log(Cab)	I	log I	log fi	fi
7	0.000001	-6	3.00E-06	-5.52288	-0.00354	0.991882
8	1.78E-06	-5.75	5.33E-06	-5.27288	-0.00472	0.989197
9	3.16E-06	-5.5	9.49E-06	-5.02288	-0.00628	0.985633
10	5.62E-06	-5.25	1.69E-05	-4.77288	-0.00837	0.980911
11	0.00001	-5	3.00E-05	-4.52288	-0.01114	0.974666
12	1.78E-05	-4.75	5.33E-05	-4.27288	-0.01483	0.966433
13	3.16E-05	-4.5	9.49E-05	-4.02288	-0.01972	0.955617

Figure 5.1. Part of the worksheet *calculations* in the workbook *activity.*

(where A7 holds the value of the electrolyte concentration) and fill down over the required range. Note the use of the named variable **MAGZb** (i.e., the magnitude of the charge on the anion **(=(Zb^2)^0.5))** to calculate the concentration of the cation.

- Calculate the log of the activity coefficient in the next column using equation (5.1) [e.g., enter a formula such as **=–0.512*Za^2*((C7^0.5/ (1+C7^0.5))–0.2*C7**, where C7 contains the value of the ionic strength calculated in the previous step] and Fill_Down over the range.
- Calculate the activity coefficient using **=10^E7,** where E7 contains the result of the previous step.
- Plot the value of the activity coefficient of the cation vs. the concentration of the electrolyte. Compare the results obtained with different classes of electrolytes (e.g., a^+/b^-, a^{2+}/b^-, a^+/b^{2-}, a^{2+}/b^{2-}). A typical set of results is shown in Figure 5.2. These clearly demonstrate that activity and concentration diverge sharply for electrolyte concentrations above about 10^{-3} M, and that the effect is more pronounced for more highly charged ions through the charge factor in equation (5.1) and because of the more rapid increase in ionic strength [equation (5.2)]. In general, the relationship be-

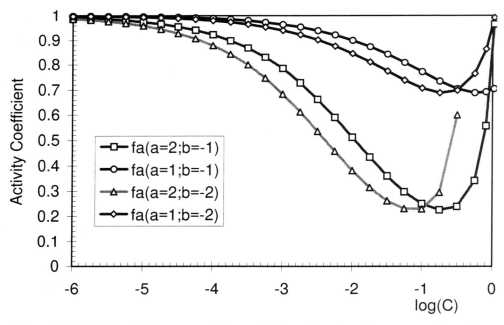

Figure 5.2. Relationship between activity coefficient of a cation, the concentration of the salt, and the type of electrolyte (a^+/b^-, a^{2+}/b^-, a^+/b^{2-}, a^{2+}/b^{2-}).

gins to break down for ionic strengths between 0.1 M and 1.0 M (note that for solutions of the type a^+b^-, the concentration is the same as the ionic strength).

A sample answer to this problem can be found in the workbook *activity*. The sheet *calculations* has the various formulas, named variables, and data required for the above calculations over the concentration range 10^{-6} M to 0.1 M. The charges on the cation and anion can be easily changed through variation of the named variables **Za** and **Zb,** respectively (see Figure 5.1). Sets of activity coefficient data for different electrolyte types have been copied into the sheet *graphs* in the same workbook to produce Figure 5.2.

5.3. KINETICS

5.3.1. Introduction

Kinetics involves the relationship between factors such as reactant concentration, temperature, and pressure that affect the velocity at which a reaction proceeds. At fixed temperature, the relationship between concentration and rate is described by a rate law. For many reactions, the rate is found to depend on the concentration raised to some power, usually 1 or 2, known as the *order.* Investigations into the order of a reaction are therefore very common. The temperature dependence of a reaction can be used to determine the *activation energy* of a reaction via the Arrhenius equation. Case studies illustrating some aspects of these quantities are given in this chapter. Equivalent examples may be found in most general texts on physical chemistry, giving data that could be analyzed in a similar manner.

Case Study 5.2. The Arrhenius Equation

For many simple reactions, the relationship between temperature and the rate constant obeys the following relationship:

$$k_2 = \ln A - (E_a/RT) \tag{5.4}$$

where R = the gas constant (8.314 J K^{-1} mol^{-1}), E_a is the activation energy (J/mol), T is the absolute temperature (K), A is the preexponential factor, and k_2 is the rate constant. A plot of k_2 vs. $1/T$ has an intercept on the y axis at $\ln A$ and a slope of $-E_a/R$. Hence both the preexponential factor and the activation energy can be found from this type of plot.

Enter the data in Table 5.1 into a spreadsheet and fit a linear regression plot for $\ln (k_2/\text{mol}^{-1} \text{ dm}^3 \text{ s}^{-1})$ vs. $1/(T/\text{K})$. Calculate the activation energy (E_a) and the preexponential factor (A) for the reaction.

Table 5.1. The Change in Rate Constant for the Decomposition of Acetaldehyde over the Temperature Range 700–850 K[1]

T/K	$k_2/\text{mol}^{-1}\,\text{dm}^3\,\text{s}^{-1}$	$1/K(\times 1000)$	$\ln(k_2)$
700	0.011	1.4286	−4.5099
730	0.035	1.3699	−3.3524
760	0.105	1.3158	−2.2538
790	0.343	1.2658	−1.07
810	0.789	1.2346	−0.237
840	2.17	1.1905	0.77473
910	20	1.0989	2.99573
1000	145	1	4.97673

[1] [1], p. 864.

5.3.2. Arrhenius Equation—Suggested Method

Experimentally, the values of the preexponential factor and activation energy for a particular reaction are determined by measuring the rate constant of the reaction at a number of temperatures and then using an Arrhenius plot [ln $(k_2/\text{mol}^{-1}\,\text{dm}^3\,\text{s}^{-1})$ vs. $1/(T/K)$. Analysis of the data yields the results $A = 1.077\times10^{12}$ and $E_a = 188.32$ kJ/mol (see Figure 5.3). The plot is shown in Figure 5.4. It is clear from the plot that the data are reasonably linear, but the bunching will mean that the estimation of the y-axis intercept will be subject to relatively large errors for

	A	B	C	D	E
1	Note:To obtain full regression report use Tools+Analysis+regression				
2					
3	T/K	$k_2$mol^{-1} dm^3 s^{-1}	1/K(x1000)	ln(k_2)	
4	700	0.011	1.428571429	-4.50986	
5	730	0.035	1.369863014	-3.35241	
6	760	0.105	1.315789474	-2.25379	
7	790	0.343	1.265822785	-1.07002	
8	810	0.789	1.234567901	-0.23699	
9	840	2.17	1.19047619	0.774727	
10	910	20	1.098901099	2.995732	
11	1000	145	1	4.976734	
12		A=	1.07906E+12		
13		Ea=	188.3182954	kJ/mol	
14					

Figure 5.3. Part of the workbook *arrhenius* showing layout of the sheet *arrhenius plot*.

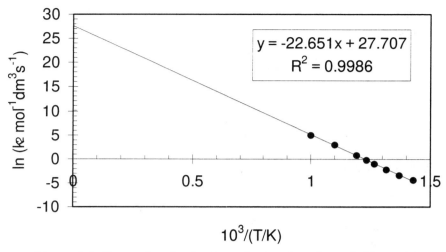

Figure 5.4. Regression plot for the *arrhenius sheet* shown in Figure 5.3.

small inaccuracies in the value of the slope, which will be further amplified through the exponential relationship between the intercept and the value of *A*. A fully worked answer to this problem can be found in the workbook *arrhenius*. Note that the **Insert_Trendline** command was used to generate the regression line, and the option to extend the trendline backwards by 1 unit used to extrapolate in order to illustrate the y-axis intercept (see section 2.6).

5.3.3. First- and Second-Order Reactions

Many chemical reactions and processes such as radioactive decay can be described using first-order kinetics, which in mathematical terms can be represented by the differential equation

$$\frac{-d[A]_t}{dt} = k[A_t] \tag{5.5}$$

where $[A]_t$ is the concentration of a substance *A* which spontaneously decomposes to form a product or products. Equation (5.5) predicts the following.

- The rate at which the substance *A* disappears is directly proportional to the concentration of A and the proportionality constant is the rate constant.
- The reaction half-life is a constant (= (ln 2)/k).

- As the concentration decreases with time, so will the rate of decrease of concentration.

The integrated form of this equation is

$$[A]_t = [A]_0 \, e^{-kt} \tag{5.6}$$

where t is the time, k is the rate constant and $[A]_0$ is the initial concentration of A.

Case Study 5.3. First-Order Kinetics

This relationship can be easily explored using Excel. Figure 5.5 shows part of the worksheet *1st order* in the workbook *kinetics* that has been set up to graph this equation. The initial concentration and rate constant have been defined as the named variables *Ao* and *kr* and their values assigned to cells D2 and D3, respectively. The time scale is defined from cell A5 using the Fill-Series command to cover the range 0–100 s, and the concentration of A calculated from cell B5 down. To achieve this, enter the formula =**Ao*EXP(-kr*A5)** in B5 and use the **Fill_Down** command to cover the time range in column A. Cells C5 down calculate the natural logarithm of the equivalent cell in column B. Cells F5 to H5 down contain the same formulas for the same time range, but with a ten-second interval rather than a one-second interval used in columns A–C. This can be useful if a reduced number of points is needed for clarity in graphing, as the use of symbols can lead to very thick displays if they are used with a

	A	B	C	D	E	F	G	H
1	First Order Kinetics		See also sheet 2nd order					
2	[A]t={A]0xe^(-kt)		Ao=	1				
3			kr=	0.1				
4	Time/s	[A]t/M(1st order)	ln[A]t			Time/s		-ln[A]t
5	0	1	0			0	1	0
6	1	0.904837418	-0.1			10	0.367879	1
7	2	0.818730753	-0.2			20	0.135335	2
8	3	0.740818221	-0.3			30	0.049787	3
9	4	0.670320046	-0.4			40	0.018316	4
10	5	0.60653066	-0.5			50	0.006738	5
11	6	0.548811636	-0.6			60	0.002479	6
12	7	0.496585304	-0.7			70	0.000912	7

Figure 5.5. Part of the workbook *kinetics* showing the named variables A_0 and **kr** (values assigned to cells D2 and D3, respectively).

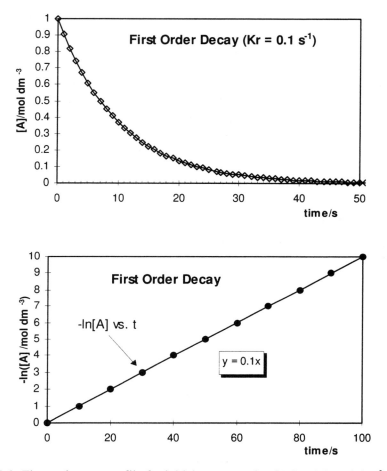

Figure 5.6. First-order rate profile for initial concentration $[A_o] = 1.0$ mol dm^{-3} and $k_r = 0.1$ s^{-1}.

large number of points. From the graph, students can estimate the half-life of the reaction (concentration of A diminished to one-half initial concentration), which, in this case, is estimated to be about 7 s (see Figure 5.6, top). Furthermore, the reaction is essentially complete after about 50 s (about seven half-lives). Exercises like this are useful for bringing home the reality behind these terms to students. Figure 5.6, bottom, shows how the value of the first-order rate constant can be estimated by graphing the natural logarithm form of equation (5.6), which can be written as

$$-\ln [A]_t = kt - \ln [A]_0 \tag{5.7}$$

The graph shows this relationship with the values 0.1 s^{-1} and 1 mol dm^{-3} used for k and $[A]_0$, respectively. The regression line is used to obtain the slope (k) and the intercept ($-\ln [A]_0$), with the latter not appearing in the Regression Equation window because it is zero in this case.

The effect of variation in rate constant value on the progress of a reaction can be easily investigated by changing the named variable **kr** in cell D3 and pasting the results into the graph. Results for **kr** = 0.1, 0.05, and 0.01 s^{-1} are shown in Figure 5.7. Clearly, a ten-fold increase in the rate constant of a first-order reaction has a major impact on the course of the reaction. In contrast to the situation when **kr** = 0.1 s^{-1} (which is essentially complete after about 50 s), if **kr** = 0.01 s^{-1}, then there is around 40% of A still unreacted after 100 s.

Case Study 5.4. Comparing First- and Second-Order Reactions

For a reaction that is second order with respect to a single reactant, the rate equation is

$$\frac{1}{[A]_t} = \frac{1}{[A]_0} + kt \tag{5.8}$$

and the half-life is given by

$$t_{1/2} = \frac{1}{k[A]_0} \tag{5.9}$$

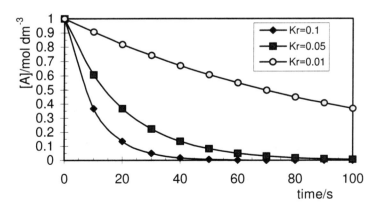

Figure 5.7. Effect of variation in kr (in s^{-1}) on reaction progress $[A]_0 = 1.0$ mol dm^{-3} in each case.

Equation (5.9) tells us that, in contrast to first-order reactions, the half-life of a second-order reaction is not a constant but a function of the reciprocal of the initial concentration of the reactant. In consequence, the half-life gets longer as the reaction proceeds, and the progress of the reaction will become considerably slower during the latter stages of a reaction in comparison to a first-order reaction with a similar rate constant. The sheet *2nd order* in the workbook *kinetics* calculates the remaining concentration of a substance A over the same timescale as the first-order reaction described above. The same named variables (**kr** and **Ao**) are used, so that changing these on the sheet *1st order* is automatically linked to the sheet *2nd order* as well. The formula **=((1/Ao)+kr*A5)^-1** (where A5 is the time) is used to calculate A_t, and this is subsequently graphed against t on the same axis as the first order values for A_t. The resulting graph (Figure 5.8) clearly shows that the second-order reaction is proceeding more slowly than the equivalent first-order reaction and that the divergence is increasing with time.

5.3.4. Suggested Extensions

- Investigate the effect of different values for **kr** and **Ao** on the reaction profiles;
- Develop a sheet to estimate the rate constant for the second order reaction from the $[A]_t$ data in a manner similar to that used for the first order reaction.

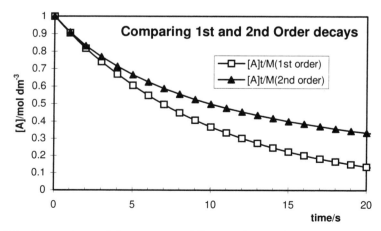

Figure 5.8. Comparison of the progress of first- and second-order reactions for a single reactant A with the values $[A]_0 = 1.0$ mol dm^{-3} and $k_r = 0.1$ used in both cases.

5.4. METAL-COMPLEX EQUILIBRIA

5.4.1. General Background

This area of physical chemistry is particularly suited to spreadsheet investigations as it is heavily mathematical and graphical in nature. Reference [2] contains many excellent examples of the use of a spreadsheet (Quattro Pro) for examining metal ion complexation that can easily be adapted for Excel. The data for the example below comes from Chapter 5 in [2]. Formation constants for many species can be obtained from any general physical chemistry textbook, or from reference texts like the *CRC Handbook of Chemistry* series, and substituted for the data in this example. In this case study, we shall look at the Cu^{2+}/NH_3 system in detail as an example of a complexation involving stepwise formation of the successive complexes. For additional examples and sample data, the reader is referred to [2].

5.4.2. Generalized Metal-Complex Equations

The formation of a complex involving a metal (M) and a ligand (L) can be represented by the generalized expression (charges omitted for simplicity):

$$M + L \rightleftharpoons ML \tag{5.10}$$

for which an equilibrium formation constant (K_f) can be written as

$$K_f = \frac{[ML]}{[M][L]} \tag{5.11}$$

A useful way of expressing the relative amounts of each form of the metal (free or complexed) is as the existing *fraction* or α of each, where

$$\alpha_M = \frac{[M]}{C_M} \quad \text{and} \quad \alpha_{ML} = \frac{[ML]}{C_M} \tag{5.12}$$

and as $[M] + [ML] = C_M$ then $\alpha_M + \alpha_{ML} = 1$, $[M] = \alpha_M/C_M$ and $[ML] = \alpha_{ML}/C_M$. Note that C_M is the total concentration of M in all its forms, $[M]$ is the concentration of the free ion, $[ML]$ is the concentration of the complex, and $[L]$ is the concentration of the free ligand.

Where stepwise formation of successive complexes is involved, a series of expressions of this type can be defined such as these for a four-step process:

$$M + L \stackrel{1}{\rightleftharpoons} ML, \quad K_{f_1} = \frac{[ML]}{[M][L]} \tag{5.13}$$

$$ML + L \overset{2}{\rightleftharpoons} ML_2, \quad K_{f2} = \frac{[ML_2]}{[ML][L]} \tag{5.14}$$

$$ML_2 + L \overset{3}{\rightleftharpoons} ML_3, \quad K_{f3} = \frac{[ML_3]}{[ML_2][L]} \tag{5.15}$$

$$ML_3 + L \overset{4}{\rightleftharpoons} ML_4, \quad K_{f4} = \frac{[ML_4]}{[ML_3][L]} \tag{5.16}$$

where K_{f_1}, K_{f_2}, K_{f_3} and K_{f_4} are the stability constants for each complex. From these, an overall formation constant (β) can be defined by combining equations (5.13) to (5.16):

$$M + nL \rightleftharpoons ML_n, \quad \beta_n = \frac{[ML_n]}{[M][L]^n}, \quad \beta_n = K_{f_1}K_{f_2}\ldots K_{fn} \tag{5.17}$$

In a similar way, formation constants can be calculated for step 1, steps 1 and 2, steps 1–3, steps 1–4, and so on. From these formation constants, we can obtain expressions for the concentration of each species containing the metal ion as follows:

$$[ML] = \beta_1[M][L]$$
$$[ML_2] = \beta_2[M][ML]^2$$
$$[ML_3] = \beta_3[M][ML]^3 \tag{5.18}$$
$$[ML_4] = \beta_4[M][ML]^4$$

Now as $C_M = [M] + [ML] + [ML_2] + [ML_3] + [ML_4]$, we can define the fraction of each species present in the following manner:

$$\alpha_M = \frac{[M]}{C_M} = \frac{1}{1 + \beta_1[L] + \beta_2[L]^2 + \beta_3[L]^3 + \beta_4[L]^4}$$

$$\alpha_{ML} = \frac{[ML]}{C_M} = \frac{\beta_1[L]}{1 + \beta_1[L] + \beta_2[L]^2 + \beta_3[L]^3 + \beta_4[L]^4}$$

$$\alpha_{ML_2} = \frac{[ML_2]}{C_M} = \frac{\beta_2[L]^2}{1 + \beta_1[L] + \beta_2[L]^2 + \beta_3[L]^3 + \beta_4[L]^4} \tag{5.19}$$

$$\alpha_{ML_3} = \frac{[ML_3]}{C_M} = \frac{\beta_3[L]^3}{1 + \beta_1[L] + \beta_2[L]^2 + \beta_3[L]^3 + \beta_4[L]^4}$$

$$\alpha_{ML_4} = \frac{[ML_4]}{C_M} = \frac{\beta_4[L]^4}{1 + \beta_1[L] + \beta_2[L]^2 + \beta_3[L]^3 + \beta_4[L]^4}$$

This gives the fraction of each species in terms of the formation constants and the free (unbound) ligand concentration. Clearly, the concentration of each Cu species can be found by multiplying the appropriate right-hand side expression in equation (5.19) by the total metal concentration.

Case Study 5.5. Investigating the Cu-NH₃ System

- Equation (5.19) shows that we can calculate the fraction of each species present from the concentration of the free ligand $[L]$, the various formation constants, and the total concentration of the metal. We shall begin therefore by defining the stability constants and total metal concentration as named variables. Figure 5.9 illustrates part of the sheet *cu-nh3 system* in the workbook *cu-nh3*. The named variables are identified in cells G2 to G6 as the total metal ion concentration and the four formation constants, respectively, and are assigned the values in the adjacent cells in column H. The values in cells H3 to H6 are in turn obtained from the log K_n values in cells F3 to F6 (for example the formula =10^F3 is placed in cell H3 and filled down over the next three cells). Defining the stability constants and

	A	B	C	D	E	F	G	H	I
1							**Named Variables**		
2					stability constants		[Cu]0	1.00E-03	
3					log kf1	4.31	kf1	2.04E+04	
4			Free Ammonia		log kf2	3.67	kf2	4.68E+03	
5					log kf3	3.04	kf3	1.10E+03	
6					log kf4	2.30	kf4	2.00E+02	
7	Log[NH3]	[NH3]	C(Cu)	[Cu]	[Cu(NH3)]	[Cu(NH3)2]	[Cu(NH3)3]	[Cu(NH3)4]	
8	-6	1.00E-06	1.00E-03	9.80E-04	2.00E-05	9.36E-08	1.03E-10	2.05E-14	
9	-5.8	1.58E-06	1.00E-03	9.68E-04	3.13E-05	2.32E-07	4.04E-10	1.28E-13	
10	-5.6	2.51E-06	1.00E-03	9.51E-04	4.88E-05	5.73E-07	1.58E-09	7.91E-13	
11	-5.4	3.98E-06	1.00E-03	9.24E-04	7.51E-05	1.40E-06	6.10E-09	4.85E-12	
12	-5.2	6.31E-06	1.00E-03	8.83E-04	1.14E-04	3.36E-06	2.32E-08	2.92E-11	
13	-5	1.00E-05	1.00E-03	8.24E-04	1.68E-04	7.87E-06	8.63E-08	1.72E-10	
14	-4.8	1.58E-05	1.00E-03	7.42E-04	2.40E-04	1.78E-05	3.09E-07	9.78E-10	

Figure 5.9. Part of the sheet *cu-nh3 system* in the workbook *cu-nh3* showing the named variables and calculation of the concentration of each Cu species.

total metal ion concentration as named variables in this manner makes variation of these quantities much easier if one wishes to explore the effect of each on the distribution of the respective forms of the complex.

- Enter appropriate titles for each column (in Figure 5.9, this is done in row 7).

- Using a logarithmic scale, enter a range of free ammonia concentrations (e.g., from log [NH$_3$] = −6 to 0 in steps of +0.2; see column A, Figure 5.9). A logarithmic scale is usual in these types of calculations due to the broad concentration ranges involved.

- Convert this to [NH$_3$] in column B by placing the formula =10^A8 in cell B8 and filling down over the range.

- Column C contains the total concentration of Cu (C_{Cu}) that will be plotted as a reference to the other curves. Use the named variable **[Cu]0** (value is in cell H2, Figure 5.9, set at 10^{-3} mol dm^{-3}) to insert this into column C so that the entire column of values can be changed instantly through changing the value in cell H2 (note that Excel does not like brackets being used in named variables, and it automatically changes the named variable in cell G2 to **Cu_0**; this is the form used in column C, e.g., =Cu_0 is placed into C8 and filled down over the range).

- Calculate the concentration of each form of Cu^{2+} for each concentration of free ammonia, beginning with [NH$_3$] = 10^{-6} mol dm^{-3}; for example, the concentration of free copper ions [Cu^{2+}] is given by the formula

=Cu_0/(1+B8*kf1+B8^2*kf1*kf2+B8^3*kf1*kf2*kf3+B8^4*kf1*kf2 *kf3*kf4),

which is placed in cell D8. Note that **B8** in this formula is the concentration of the free ligand [*L*], **Cu_0** is the total concentration of metal ion (C_M) and **kf1, kf2, kf3,** and **kf4**[†] are the stability constants as described in equations (5.13) to (5.16). To calculate the concentration of the [Cu(NH$_3$)]$^{2+}$ complex, the formula =D8*kf1*B8 is placed in E8, and likewise the formulas **=D8*kf1*kf2*B8^2, =D8*kf1*kf2*kf3*B8^3,** and **=D8*kf1*kf2*kf3*kf4*B8^4** are entered into cells F8, G8, and H8, for the complexes [Cu(NH$_3$)$_2$]$^{2+}$, [Cu(NH$_3$)$_3$]$^{2+}$, and [Cu(NH$_3$)$_4$]$^{2+}$, respectively. These are then filled down over the range of free ligand concentration investigated.

- Graph the data using the scatterplot option, with the free ligand concentration (column B) as the x-axis data. Don't forget to pick up the title row (7 in this example) for inclusion in the legend. The result should be similar to Figure 5.10.

[†]Avoid using variable identifiers such as B1, B2 etc. as Excel will treat these as cell references!

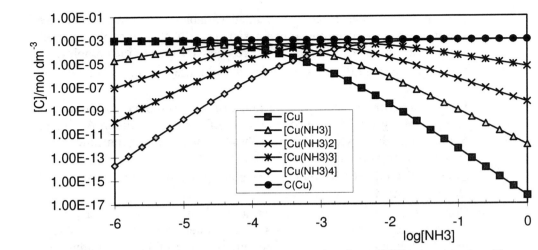

Figure 5.10. Distribution of each Cu species as a function of NH_3 concentration. Note that $C(Cu)$ is the total amount of copper, and $[Cu]$ is the concentration of free (uncomplexed) copper ions.

This figure clearly demonstrates that at low free ligand concentrations ($[NH_3] < 10^{-5}$ mol dm^{-3}), the copper exists predominantly as free copper ions, Cu^{2+}. However, the free ion concentration drops sharply above about $[NH_3] \approx 10^{-4}$ mol dm^{-3} and the various complexed forms begin to predominate. Above $[NH_3] \approx 10^{-2}$ mol dm^{-3}, the most highly complexed form, $[Cu(NH_3)_4]$,$^{2+}$ begins to dominate.

While this is an informative and useful study, in practice we are interested in the distribution of the various species as a function of the total ion concentration and the total ligand concentration, which are usually the known parameters in an experimental situation. This can be achieved using the data calculated above. The total ligand concentration (C_{NH_3}) can be obtained from

$$C_{NH_3} = [NH_3] + [Cu(NH_3)]^{2+} + 2[Cu(NH_3)_2]^{2+}$$
$$+ 3[Cu(NH_3)_3]^{2+} + 4[Cu(NH_3)_4]^{2+} \qquad (5.20)$$

This has been calculated in cell J8 using the formula **=B8+E8+2*F8+3*G8+ 4*H8.** The logarithm of this is calculated in K8 and the formulas filled down over the range of the previous study. To simplify the graphing operation, the distribution of each species is copied into columns M–Q from the equivalent cells in columns D–H (see Figure 5.11). Use the **Paste_Special** command and the **Paste_Link** option to link the values in these cells, so that if a stability con-

	C_{NH3}	log(C_{NH3})	C(Cu)	[Cu]	[Cu(NH3)]	[Cu(NH3)2	[Cu(NH3)3	[Cu(NH3)4]
				Total Ammonia				
8	2.119E-05	-4.67378	0.001	0.00098	2E-05	9.36E-08	1.03E-10	2.05E-14
9	3.339E-05	-4.4764	0.001	0.000968	3.13E-05	2.32E-07	4.04E-10	1.28E-13
10	5.242E-05	-4.28052	0.001	0.000951	4.88E-05	5.73E-07	1.58E-09	7.91E-13
11	8.186E-05	-4.08692	0.001	0.000924	7.51E-05	1.4E-06	6.1E-09	4.85E-12
12	0.0001268	-3.89678	0.001	0.000883	0.000114	3.36E-06	2.32E-08	2.92E-11
13	0.0001942	-3.71175	0.001	0.000824	0.000168	7.87E-06	8.63E-08	1.72E-10
14	0.0002924	-3.53398	0.001	0.000742	0.00024	1.78E-05	3.09E-07	9.78E-10

Figure 5.11. Arrangement of part of the sheet *cu-nh3 system* for calculating and plotting distribution of Cu species as a function of total NH_3 concentration (C_{NH3}).

stant or the total metal concentration is changed, then both sets of values are affected. A scattergraph of **log(C_{NH3})** vs. the various Cu^{2+} species is shown in Figure 5.12. This enables the concentration of each species to be quickly estimated if the total metal and ligand concentrations are known. Obviously, this type of spreadsheet provides an ideal environment for studying speciation. For example, the total metal concentration can be varied, the number of complexation steps varied (set the appropriate stability constants to zero beginning with the highest order), K_{f4}, or the relative stability constant values varied to explore the limiting situations for ensuring that a particular complex predominates in the solution.

For further examples and useful discussion on this topic, the reader is referred to [2].

5.5. TITRATION CURVES

The calculation of titration curves is another area that lends itself to the spreadsheet approach. In Chapter 8, we shall look at strong acid–strong base curves and show how an Excel spreadsheet can be provided with a simple user interface for inputing data. In this section, we shall concentrate on weak acid–strong base curves and derive the equations used to calculate the titration curves.

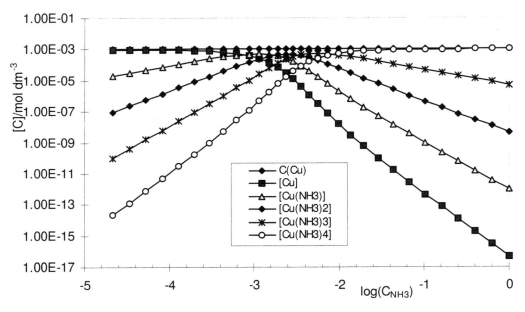

Figure 5.12. Plot of the distribution of the various Cu^{2+} species as a function of total NH_3 concentration.

5.5.1. Introduction

This is a popular topic in undergraduate chemistry studies and it lends itself to spreadsheet investigations because of its graphical nature. In this section, we will restrict ourselves to the titration of a weak acid by a strong base (NaOH). Freiser [2] gives a useful summary of the theory of titrations and acid–base equilibrium generally and also gives useful routines for assembling theoretical curves from the equations. We shall build on his approach and develop a more flexible spreadsheet that limits the pH range of the calculations to realistic values only.

5.5.2. Derivation of Equations Used

In general, the dissociation of a weak, monoprotic acid (HA) can be represented by

$$HA \rightleftharpoons H^+ + A^-; \quad K_a = \frac{[H^+][A^-]}{[HA]} \tag{5.21}$$

where A = the conjugate base of the weak acid, and K_a is the acid dissociation constant. Now the total concentration of acid (C_A) is related to the amount of associated and dissociated forms of the acid as

$$C_A = [HA] + [A^-] = \alpha_0 C_A + \alpha_1 C_A \qquad (5.22)$$

where α_0 and α_1 represent the fraction of the associated and dissociated forms of the acid, respectively ($\alpha_0 + \alpha_1 = 1$).

Combining equations (5.21) and (5.22), we get for [HA] and [A$^-$]

$$[HA] = C_A \left(\frac{[H^+]}{[H^+] + K_a} \right), \quad [A^-] = C_A \left(\frac{K_a}{[H^+] + K_a} \right) \qquad (5.23)$$

and substituting for the fraction of each form,

$$\alpha_0 = \left(\frac{[H^+]}{[H^+] + K_a} \right), \quad \alpha_1 = \left(\frac{K_a}{[H^+] + K_a} \right) \qquad (5.24)$$

During the titration, a charge balance must be obeyed. In the case of the addition of NaOH to HA, the expression for the charge balance is

$$[Na^+] + [H^+] = [OH^-] + [A^-] \qquad (5.25)$$

At any point in the titration, the concentration of sodium ions can be calculated from the number of moles divided by the total concentration

$$[Na^+] = \frac{V_B C_B}{V_A + V_B}, \qquad (5.26)$$

where V stands for volume in mL, C is the total concentration, and the subscripts A, B represent the acid and base, respectively.

Likewise, the concentration of dissociated acid can be calculated from the total acid concentration and the fraction of acid existing in the dissociated form

$$[A^-] = \left(\frac{V_A C_A}{V_A + V_B} \right) \alpha_1 \qquad (5.27)$$

Substituting equations (5.27) and (5.26) into equation (5.25) we obtain

$$\left(\frac{V_B C_B}{V_A + V_B} \right) + [H^+] = [OH^-] + \left(\frac{V_A C_A}{V_A + V_B} \right) \alpha_1 \qquad (5.28)$$

and solving for V_B we arrive at

$$V_B = V_A \left(\frac{C_A \alpha_1 - [H^+] + [OH^-]}{C_B + [H^+] - [OH^-]} \right) \tag{5.29}$$

This equation enables us to calculate the volume of base added when the titration is at any pH. Assuming $[H^+]$ is known, then $[OH^-]$ can be obtained from $pOH = 14 - pH$. As C_A and C_B are known and α_1 can be easily calculated from equation (5.27), the volume of base can be calculated over the pH range of the titration, and the two quantities graphed against each other to give the titration curve. It is a good idea to estimate the pH range of the titration and to limit the values to this range, rather than assume the range 0–14; otherwise, changing dissociation constants and concentrations can lead to very strange curves, with infinitely large base volumes at the end and negative base volumes at the start!

Case Study 5.6. Weak Acid–Strong Base Titration Curves

Suggested Method (see workbook ethanoic and Figure 5.13)

- Define named variables for V_A, C_A, C_B, and K_a (defined in cells A2–A5, with values placed in cells B2–B5.
- Estimate the initial pH of the acid sample from the approximation:

$$pH_{min} = -\log(\sqrt{K_a C_A}) \tag{5.30}$$

	A	B	C	D	E	F	G	H	I	J
1	Named Variables				Constants					
2	Va	10		pH(max)	12.9					
3	Ca	0.1		pH(min)	2.88					
4	Cb	0.1		inc	0.077077					
5	Ka1	1.74E-05		pKas	4.76					
6	scale	0.02		changes derivative scale						
7	offset	0.5		offsets derivative						
8										
9			Titration Curves for Weak Acid - Strong Base						Gran's Plot	
10	Vb	d(pH)/d(V)	pH	[H]	pOH	[OH]	alpha1			
11	-0.00169		2.88	0.001318	11.12	7.59E-12	0.013011		Vb	Vb*10^(-pH)
12	0.044111	0.543664	2.96	0.001104	11.04292	9.06E-12	0.015499		-0.00169	-2.2E-06
13	0.091249	0.542429	3.03	0.000924	10.96585	1.08E-11	0.018453		0.044111	4.87E-05
14	0.141081	0.540135	3.11	0.000774	10.88877	1.29E-11	0.021958		0.091249	8.43E-05

Figure 5.13. Layout of the sheet *ethanoic acid* in the workbook *ethanoic* for construction of weak acid–strong base titration curves and Gran's plots.

which is valid as long as K_a is not too large or the acid too dilute. Limit the initial pH of the curve to this value by entering this value in the first cell of the pH range. In the workbook *ethanoic* this formula is placed in cell E3 [=−LOG((Ka1*Ca)^0.5)+0.01], and its value linked to the named variable **pH(min)**. The cell C11 is also linked to this value using the formula =pH(min) (as mentioned previously, Excel does not like parentheses in named variables, and the name is automatically changed to **pH_min,** but this does not affect the calculation). The correction factor +0.01 at the end of the formula is to prevent the volume of base going slightly negative due to the limitations of the approximation.

- Likewise, limit the maximum of the *p*H range to a realistic value. For a symmetrical curve, we can assume that, for a monobasic–monoacidic system, the amount of base added must be twice the initial number of moles of acid present. Half the base added is neutralized by the end of the titration, and the total volume can be calculated to give [OH⁻] in excess, which enables us to calculate the final *p*H. The resulting formula is =14+LOG(Ca*Cb/(2*Ca+Cb)) which is entered in cell E2 and its contents linked to the final cell in the pH range (C141).

- An increment can now be calculated, which is added sequentially to each cell in the range. This value is calculated in cell E4 using the formula =(E2 − E3)/130 (there are 130 cells in the range). This value is linked to the named variable **inc** in cell D4.

- Enter the range of *p*H values. This begins with cell C11 (set to **pH_min**). Use =C11+inc to enter the next value of the series, and then fill this equation down over the rest of the series (i.e., over the cell range C12:C141).

- Now calculate [H⁺], *p*OH, [OH⁻], and α_1 [the latter using the formula =Ka1/D11+Ka1); see equation (5.24)]. The values are calculated in cells D11 to G11, respectively.

- The volume of base required to obtain that particular *p*H is calculated via equation (5.29). In the workbook *ethanoic* it is entered as the formula =Va*(Ca*G11-D11+F11)/(Cb+D11-F11) in cell A11. The formulas can be filled down over the range of *p*H values to give the base volumes required for each pH value, once the values for the acid concentration, volume, dissociation constant, and the base concentration are entered in the appropriate cells. A scattergraph of A10:A141 vs. C10:C141 (Figure 5.14) shows the well-known weak-acid–strong base curve for ethanoic acid vs. NaOH for the starting conditions $C_a = 0.1$ mol dm⁻³, $V_a = 10$ mL, $C_b = 0.1$ M, $pK_a = 4.76$. (The pK_a can be entered in cell E5—the sheet automatically converts this into the K_a in cell B5.

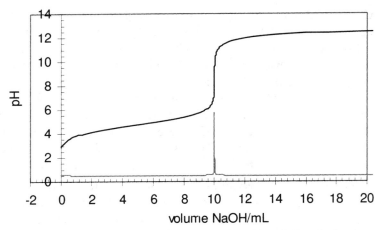

Figure 5.14. Theoretical titration curve for ethanoic acid and first derivative vs. NaOH under the conditions shown in Figure 5.13 (pK_a = 4.76, initial volume of acid = 10 mL; concentration of acid = concentration of base = 0.1 mol dm^{-3}.

There are many ways to assemble a weak acid–strong base titration curve, each with varying degrees of flexibility. This version can cope with a wide range of pKa values and with various concentrations of acid or base, or variable acid sample volume. It allows the relationship between equivalence point and these variables to be investigated, and perhaps more importantly, the relationship between the size of the inflection at equivalence and K_a can be probed. Figure 5.16 shows the effect of changing the pK_a of the weak acid to 9.0. Clearly the inflection is drastically reduced in size, and accurate determination of the equivalence point becomes more difficult. The suitability of various indicators for signaling the equivalence point can easily be determined from a knowledge of the pK_a of the indicator and the fact that the color change is essentially complete by plus or minus 1 pH unit of the indicator pK_a.

5.5.3. The Derivative Curve

As mentioned, when the pK_a of the acid becomes large, location of the equivalence point becomes more difficult. Differentiation is a well-known method for enhancing features in graphs, as the first derivative is basically a plot of the change in slope vs. the x ordinate.

In spreadsheets, an approximate first derivative of a curve can be obtained by simple subtraction of sequential cell values, which is very effective providing the number of cells is large (and hence the increment is acceptably small). This technique and the further use of a second derivative, are common methods for equivalence point determination in autotitrator software. In this example, we shall look only at the first derivative, but obviously the second derivative can easily be obtained through subtraction of the first derivative values.

Example

Method

In the workbook *ethanoic* the first derivative is calculated in column B beginning with cell B12, which contains the formula **=(C12–C11)/(A12–A11)*scale+offset.** This can be seen to represent the first derivative $d(pH)/d(V)$, along with two additional factors, *scale* and *offset,* which enable the derivative plot to be magnified or attenuated and offset from the x axis. The values of scale and offset can be easily changed through the equivalent named variables in cells A6 and A7, whose values are defined in cells B6 and B7, respectively. Figure 5.14 shows the first derivative plot $[d(pH)/d(V)$ vs. $V]$ for ethanoic acid vs. NaOH (pK_a = 4.76, initial volume of acid = 10 mL; concentration of acid = concentration of base = 0.1 mol dm^{-3}). This illustrates how useful the first derivative is at locating the equivalence point accurately. Clearly, the change in pH around the equivalence point is very sharp for this titration and can be located very accurately. In addition, the first derivative also identifies the buffer region of the titration during which both the weak acid and its salt exist, and the pH changes by a very small amount while the base is being added (between pH 4 and 5). In fact, the inverse of the first derivative actually quantifies the volume of base required to bring about a unit change in the pH (the buffer capacity). This can also easily be calculated and plotted.

Figure 5.16(a) shows the same type of plot for a weak acid of pK_a = 9.0. In comparison with the previous case, the derivative curve shows that the sharpness at the inflection is greatly reduced, making it difficult to locate precisely using a visual indicator. Hence these titrations are normally performed with an autotitrator, using a glass electrode to follow the changing pH and a first or second derivative to locate accurately the equivalence point. The buffer region is also affected as the pH changes significantly during the initial part of the titration and as the equivalence point is approached.

5.6. GRAN'S PLOT

5.6.1. Advantages of Gran's Plot

The Gran's plot is an alternative method for locating the equivalence point of a titration that has several important advantages, listed below, over the conventional titration curve method.

1.   It is a linear plot, and least-squares regression analysis can be employed to determine the best-fit slope (from which the acid dissociation constant is obtained) and x-axis intercept (i.e., volume of base added at equivalence) with great accuracy (as many points can be used to determine the regression parameters in contrast to the titration curve where only one point is used).

2.   Statistical errors for the volume of base required can be calculated from the regression line.

3.   The experimental data can be taken well before the equivalence point and extrapolated to locate the equivalence point, which saves time, reagent consumption and improves accuracy, as measurements around the equivalence point are prone to experimental error, due to the more rapid change in pH for small additions of base.

5.6.2. Background

Before the equivalence point, the fraction of remaining acid to neutralized acid (f) is given by

$$f = \frac{[\text{HA}]}{[\text{A}^-]} = \left(\frac{V_E - V_B}{V_B} \right) \tag{5.31}$$

where V_E is the volume of base added at equivalence and it is assumed that the acid exists mainly in the protonated form HA, which is valid for weak acids that are not very dilute. Substituting into equation (5.21) we get

$$K_a = [\text{H}^+] \left(\frac{V_B}{V_E - V_B} \right) \tag{5.32}$$

which can be rearranged to give;

$$[\text{H}^+]V_B = K_a(V_E - V_B) \tag{5.33}$$

Hence, a plot of $[\text{H}^+]V_B$ vs. V_B will be linear with an x-axis intercept at $V_E = V_B$, a slope of $-K_a$, and a y-axis intercept of $K_a V_E$.

Figure (5.15) and figure 5.16(b) shows Gran's plots of the titration curves in figure (5.14) and (5.16a), respectively, using data from well before the equivalence point, and the **Insert_Trendline** option to obtain the regression line. The trendline can extrapolated via the Options dialog box, which also allows the user to display the regression equation and correlation coefficient on the graph. In each case, the K_a is obtained from the slope of the regression equation (1.72 × 10^{-5} and 1.00×10^{-9} dm^3 mol^{-1}, respectively), and the y-axis intercept gives K_a

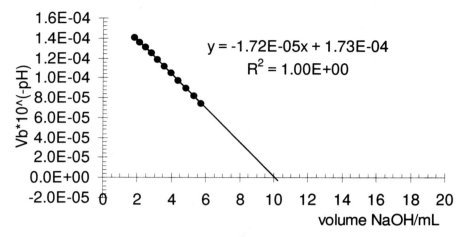

Figure 5.15. Gran's Plot of part of the data calculated for Figure 5.14 illustrating the use of regression analysis to find the x-axis intercept (reaction equivalence point). Note that the regression points used to calculate the intercept are located well before the equivalence point and that the slope of the regression line gives an estimation of the acid K_a, while the correlation coefficient gives an indication of the data quality. Note also that regression analysis will also give an estimate of the error of the intercept and of the slope (not shown).

V_E. Note that when the acid–base parameters (pK_a, C_A, C_B, and V_A) are varied, the range of points selected for the Gran's plot will have to be adjusted for optimum results.

5.7. TITRATIONS INVOLVING POLYBASIC ACIDS

Where more than one proton is available per mole of acid, the above approach can be easily extended. For example, the dissociation of phosphoric acid can be described by the following expressions:

$$H_3A \rightleftharpoons H_2A^- + H^+, \quad K_{a1} = \frac{[H_2A^-][H^+]}{[H_3A]}$$

$$H_2A^- \rightleftharpoons HA^{2-} + H^+, \quad K_{a2} = \frac{[HA^{2-}][H^+]}{[H_2A^-]} \qquad (5.34)$$

$$HA^{2-} \rightleftharpoons A^{3-} + H^+, \quad K_{a3} = \frac{[A^{3-}][H^+]}{[HA^{2-}]}$$

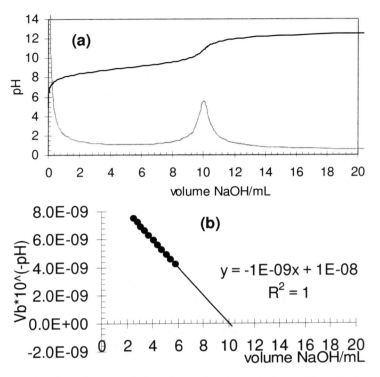

Figure 5.16. Titration Curve and first derivative (a) and Gran's Plot (b) for weak acid ($pK_a = 9$), concentration of acid = 0.1 mol dm^{-3}, volume acid = 10 mL titrated with 0.1 M strong base.

where $A = PO_4^{3-}$.

The fraction of acid existing in each form can be related via the following equation:

$$C_A = [H_3A] + [H_2A^-] + [HA^{2-}] + [A^{3-}] = \alpha_0 C_A + \alpha_1 C_A + \alpha_2 C_A + \alpha_3 C_A \quad (5.35)$$

where C_A is the total concentration of the acid in all forms and the subscript on each α denotes the number of protons removed.

From this expression, we can state

$$\alpha_0 = \frac{[H_3A]}{C_A}, \quad \alpha_1 = \frac{[H_2A^-]}{C_A}, \quad \alpha_2 = \frac{[HA^{2-}]}{C_A}, \quad \alpha_3 = \frac{[A^{3-}]}{C_A} \quad (5.36)$$

and substituting into the equilibrium expressions [equation (5.34)],

$$K_{a_1} = \frac{[H^+]\alpha_1}{\alpha_0}$$

$$K_{a_2} = \frac{[H^+]\alpha_2}{\alpha_1} \tag{5.37}$$

$$K_{a_3} = \frac{[H^+]\alpha_3}{\alpha_2}$$

We can now derive expressions for each fraction in terms of α_0, the concentration of hydrogen ions and the dissociation constants,

$$\alpha_1 = \frac{\alpha_0 K_{a1}}{[H^+]}$$

$$\alpha_2 = \frac{\alpha_1 K_{a2}}{[H^+]} = \frac{\alpha_0 K_{a1} K_{a2}}{[H^+]^2} \tag{5.38}$$

$$\alpha_3 = \frac{\alpha_2 K_{a3}}{[H^+]} = \frac{\alpha_0 K_{a1} K_{a2} K_{a3}}{[H^+]^3}$$

and remembering that the sum of all fractions present equals 1, we can write

$$1 = \alpha_0 + \frac{\alpha_0 K_{a1}}{[H^+]} + \frac{\alpha_0 K_{a1} K_{a2}}{[H^+]^2} + \frac{\alpha_0 K_{a1} K_{a2} K_{a3}}{[H^+]^3} \tag{5.39}$$

which we can rearrange in terms of α_0,

$$\alpha_0 = \frac{[H^+]^3}{[H^+]^3 + [H^+]^2 K_{a1} + [H^+] K_{a1} K_{a2} + K_{a1} K_{a2} K_{a3}} \tag{5.40}$$

Representing the denominator of equation (5.40) as D, we can similarly arrive at expressions for the fractions of other species present by substituting equation (5.40) for α_0 in the various expressions in equation (5.38) to arrive at

$$\alpha_1 = \frac{[H^+]^2 K_{a1}}{D}, \qquad \alpha_2 = \frac{[H^+] K_{a1} K_{a2}}{D}, \qquad \alpha_3 = \frac{K_{a1} K_{a2} K_{a3}}{D} \tag{5.41}$$

From these expressions, we can calculate the fraction of each form of the acid present at any pH from the acid dissociation constants.

A more generalized form of equation (5.29) enables us to calculate the volume of base added at any pH during the titration:

$$V_B = V_A \left(\frac{C_A \sum n a_n - [H^+] + [OH^-]}{C_B + [H^+] - [OH^-]} \right) \tag{5.42}$$

where the n in the summation refers to the number of protons removed from the acid. This expression enables us to relate the volume of base added to the pH at any point during the titration, and, hence, the titration curve can be calculated in the same way as for the weak acid–strong base curve described in Case Study 5.6. One method is set out in the workbook *h3po4*, part of which is illustrated in Figure 5.17. Named variables are defined (**Va, Ca, Cb, Ka1, Ka2,** and **Ka3**) and **pH(max), pH(min)** and **inc,** which are estimated in a manner similar to that described in Case Study 5.6. Using the limits defined by pH(min) and pH(max), a pH range is defined starting in cell B11 which is set at pH(min), and incremented over the range of cells B12:B141 using a step value of *inc* (use the formula =**B11+inc** in cell B12 and fill B12 down). [H$^+$] is calculated in column D, the pOH and [OH$^-$] in columns E and F, respectively, the fractions of each form present in columns G–J and the factor $\sum n\alpha_n$ in column K, beginning with

	A	B	C	D	E	F	G	H	I	J	K	L
1				**Named Variables**								
2			Va	10	pH(max)	12.3979						
3			Ca	0.1	pH(min)	1.675						
4			Cb	0.1	inc	0.08248						
5			Ka1	0.00708	pKa1	2.15						
6			Ka2	6.3E-08	pKa2	7.2						
7			Ka3	4E-13	pKa3	12.4						
8												
9			**Titration Curves for Phosphoric Acid**									
10	Vb	pH	d(pH)/d(v [H]		pOH	[OH]	alpha3	alpha2	alpha1	alpha0	sum(ialp	sum a
11	0.32666	1.675		0.02113	12.325	4.7E-13	1.4E-17	7.5E-07	0.25092	0.74908	0.25092	
12	0.96598	1.75748	0.15642	0.01748	12.2425	5.7E-13	2.4E-17	1E-06	0.28827	0.71173	0.28827	
13	1.60928	1.83997	0.15545	0.01446	12.16	6.9E-13	4E-17	1.4E-06	0.32874	0.67126	0.32875	
14	2.25433	1.92245	0.15503	0.01195	12.0775	8.4E-13	6.5E-17	2E-06	0.37193	0.62807	0.37193	
15	2.8975	2.00494	0.15548	0.00989	11.9951	1E-12	1.1E-16	2.7E-06	0.41726	0.58274	0.41727	
16	3.53382	2.08742	0.15715	0.00818	11.9126	1.2E-12	1.7E-16	3.6E-06	0.46404	0.53596	0.46404	
17	4.15727	2.1699	0.1604	0.00676	11.8301	1.5E-12	2.8E-16	4.8E-06	0.51145	0.48854	0.51146	
18	4.76124	2.25239	0.16557	0.00559	11.7476	1.8E-12	4.5E-16	6.3E-06	0.55866	0.44133	0.55868	
19	5.33909	2.33487	0.17306	0.00463	11.6651	2.2E-12	7.1E-16	8.3E-06	0.60484	0.39515	0.60485	

Figure 5.17. Part of the workbook *h3po4* showing the named variables.

=(3*G11+2*H11+I11) in cell K11. The accuracy of the fraction calculations is checked in column L, which is the simple summation of all fractions present (=1 if calculations are correct).

Figure 5.18 shows the curve obtained using the acid dissociation constants for phosphoric acid [pK_a (1–3) = 2.15, 7.20, and 12.4, respectively], $C_A = C_B$, and V_A = 10 mL. The first two equivalence points are clearly identified in the titration curve and the first derivative curves. The third equivalence point for the removal of the third proton from the acid does not appear, as the HPO_4^{2-} species is too weak an acid to be deprotonated by NaOH, or alternatively, HPO_4^{2-} is a base of too similar strength to NaOH. Figure 5.19 is a plot of the fraction of each form of the acid present during the course of the titration shown in Figure 5.18. At the beginning of the titration, the phosphoric acid exists mainly as H_3PO_4 (ca. 75%) and $H_2PO_4^-$ (ca. 25%). At the first equivalence point, (10 mL NaOH added), the acid is almost 100% $H_2PO_4^-$, and at the second equivalence point (20 mL NaOH added), it is almost 100% HPO_4^{2-}. With further addition of base, the final proton is only partially removed (note the curvature of this portion of the graph), and both HPO_4^{2-} and PO_4^{3-} coexist in equilibrium. This type of graph is of great utility in understanding the relationship between stepwise reactions and the extent of reaction.

For example, Figure 5.20 shows the result obtained using pK_a values of 2.5, 6.0, and 9.0 in place of the phosphoric acid values. The difference of about three units is enough to ensure that each form of the acid is essentially titrated in turn,

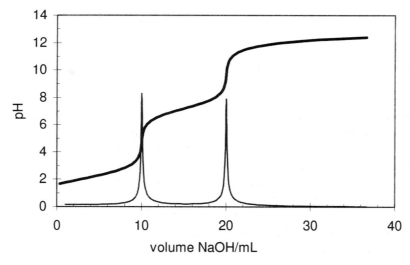

Figure 5.18. A plot of the titration of H3PO4 by NaOH and the first derivative.

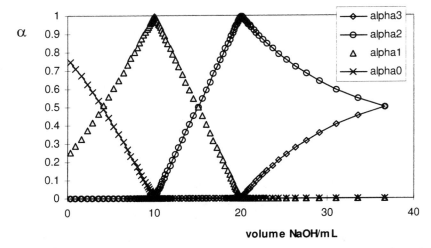

Figure 5.19. Plot of the fraction of each species present during the course of the titration shown in Figure 5.18.

provided a strong enough base is used to ensure the third proton is removed from the acid. This is confirmed by the plot of the fraction of each species present during the course of the titration, which shows that each proton is almost completely titrated in turn. The plot shows that the third proton (alpha3) begins to titrate before the second is complete (alpha2 reaches about 0.95 after 20 mL NaOH added). This explains why the second inflection is the least sharp in the first derivative curve. This point is further explored in Figure 5.21, which shows the effect of using pK_a values of 5, 7, and 11. The fraction of each species plot (bottom) shows that the acid exists almost completely as the H_3A form before the addition of any base (very little dissociation in water due to the high pK_{a1} value). After addition of 10 mL of base, H_3A, H_2A^-, and HA^{2-} coexist in equilibrium. However, after addition of 20 mL of NaOH, these forms of the acid are almost completely converted to the HA^{2-} form. Further addition of base does not completely remove the final proton, since the final pH is not sufficiently higher than the pK_a3 value. This explains why the titration curve shows only one significant inflection, at the point of almost complete removal of the second proton after 20 mL of base is added. Using this approach, students can quickly grasp the importance of having a difference of around three units in successive pK_a values, if each species is titrated separately, and giving independent inflection points.

This point is nicely reinforced in Figure 5.22, which shows the titration of citric acid with NaOH. The graphs were produced by simply substituting the three

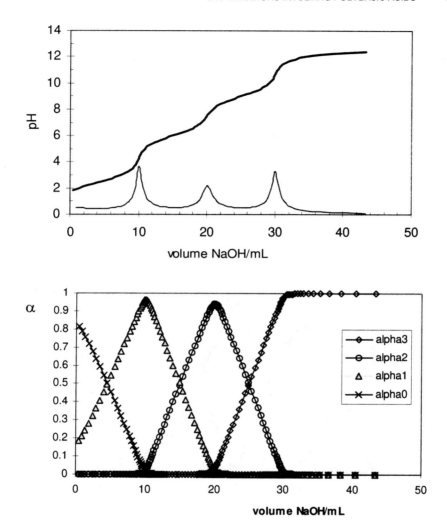

Figure 5.20. Multiple inflections in the titration of a triprotic acid with NaOH showing the titration curve and first derivative (top) and the fraction of each species present (bottom) during the course of the titration.

pK_a values of the triprotic citric acid (3.13, 4.77, and 6.4) in place of those of phosphoric acid. The fraction of each species plot (bottom) clearly shows that only the last proton is titrated to completion, while the other two overlap (which is to be expected from the small differences in pK_{a1}, pK_{a2}, and pK_{a3}, and the relatively low value of pK_{a3}, which is well below the final pH of the titration solution

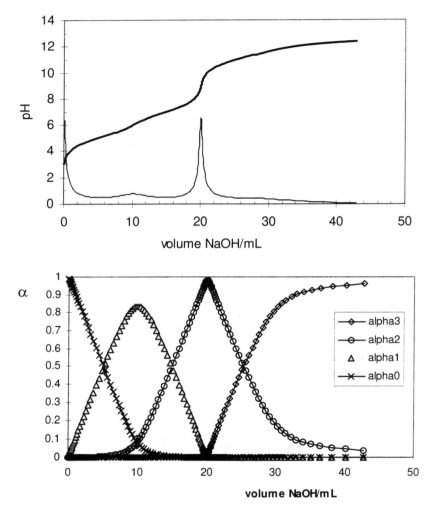

Figure 5.21. Effect of using *pKa* values of 5, 7, and 11 for the titration of a triprotic acid with NaOH.

in excess NaOH, ensuring that it is efficiently titrated by the base. Obviously, similar workbooks for the titration of diprotic acids can be prepared using the same method to develop the various equations for each fraction present, as can workbooks for the titration of weak bases by strong acid. These would make useful assignments to test whether the details in this section have been fully understood. There's nothing like trying to find out if you can do it!

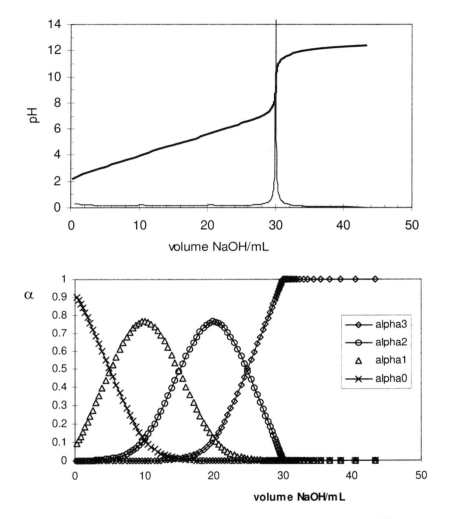

Figure 5.22. Titration of citric acid (0.1M, 10 mL) with 0.1 M NaOH and first derivative plot (top), and fraction of each species present (bottom).

5.8. CONCLUSION

Titration curves and chemical equilibria in general provide a very rich source of material for which the spreadsheet approach is ideal in that it involves fairly complex mathematical relationships and it benefits greatly from graphical analysis. In Excel, a customized interactive environment can be developed in which the relationship between chemical properties such as equilibrium constants and real-world behavior can be explored in an exciting, dynamic manner.

REFERENCES

[1] Atkins, P. W., *Physical Chemistry,* 1st ed. Oxford University Press, 1978.

[2] Freiser, Henry, *Concepts and Calculations in Analytical Chemistry—Spreadsheet Approach,* rev. ed. CRC Press, Boca Raton, FL, 1992.

[3] Parker, O. J. and Breneman, G. L., *Spreadsheet Chemistry.* Prentice-Hall, NJ. 1991.

CHAPTER 6

IMPORTING AND PROCESSING INSTRUMENTAL DATA

6.1. INTRODUCTION

Excel can provide a very useful and flexible means both for customizing the graphical presentation of instrumental data and for applying many types of data processing options. Unlike the "black-box" approach that is all too common these days and that does not encourage an understanding of the background to the myriad of data processing options, the customization of Excel for a particular task, while somewhat time-consuming for the novice, is well worth the effort. Because the user has complete control over how the data are processed, there is a certain reassurance that nothing is happening to the data of which the user is unaware. Additionally, the flexibility arises from the ability of the user to enter his or her own algorithms and formulas, rather than being restricted to the range offered by a particular package.

However, there are limitations to what Excel can do, mainly in the size of the data arrays (4000 value limit), the speed at which complex calculations can be performed (particularly if there is a graphical output), and the complexity of the operations (for more complex options one is better off using a dedicated signal processing or chemometrics package).

In this chapter we will explore some examples of using Excel as a flexible platform for processing and displaying experimental data and look in detail at how it can be applied to the teaching of some aspects of digital signal processing (DSP).

6.2. IMPORTING INSTRUMENTAL DATA

The first task to overcome in this area is how to get the instrumental data into Excel. Instrumental data sources can vary from spectrometers and sensors linked to data acquisition cards to chromatographs and potentiostats. These days, almost every laboratory instrument is linked to a PC, which will be running either Windows or DOS compatible software. There are several methods for getting the instrumental data into the Excel environment.

1. If the instrument is running Windows compatible software, it may be possible to *cut* and *paste* the data into Excel via the Clipboard. In order to do this, the package must normally allow access to a spreadsheet-type display of the experimental data.
2. If the above option is not possible, or if the instrument software is running under DOS (a surprising number still do), then try to save the data in ASCII or TEXT format. This is usually possible under the *save* or *export* command menus for almost all packages. Importing TEXT files into Excel is relatively easy because the package walks the user through the various steps. Examples in this chapter illustrate the process.
3. In some cases, an instrument may have a noncompatible operating system. These present a problem in terms of data transfer, but on occasion, an RS-232 or GPIB port may be available through which the data can be extracted into a PC. One of the case studies in this chapter demonstrates how this problem was overcome for a particular instrument.

6.3. IMPORTING TEXT (ASCII) FILES

IRTEXT.txt is a sample text file obtained from a DOS-based IR spectrophotometer. Importing the data into Excel is facilitated by the Text Import Wizard, which is automatically implemented on opening a text file. The first task is to find the file on the disk. Make sure that the File View window is set to list all files using the DOS wildcard (*.*) or by selecting *all files* option in the view window. This is important as the package defaults to Excel format files and text files may not appear in the Filename View window. When the file is opened, Excel walks the user through three stages, which allows control of the manner in which the data are arranged in Excel. This is a major improvement on previous versions, in which the data often had to be manually rearranged on importation. The operations involved to import an ASCII file (IRTEXT.TXT) are as follows.

- Select *File+Open* from the command menu bar and select *all files* in the

View window. Double click on the filename to open up the file. The first of three windows of the Text Import Wizard will appear (Figure 6.1).

- This window allows the user to check whether the text file contains the relevant information. Click on the **Next** option to proceed to the second stage.
- The second stage (Figure 6.2) allows the user to select the data delimiter so that data series can be separated into columns. Text data are often arranged in arrays that have a delimiter character (usually a *tab, semicolon, comma,* or *space*) to separate them into columns. Excel allows the user to select the delimiter and to observe the effect of each delimiter on the data, with the columns being separated by vertical black bars when the appropriate delimiter is selected (in this case *tab*). Select **Next** to move to the final stage of the process.
- The third stage (Figure 6.3) allows further options, which can be applied following successful separation of the data into columns and also allow certain columns to be skipped from the importation process if the data is of no relevance. Select **Finish** to end the process.

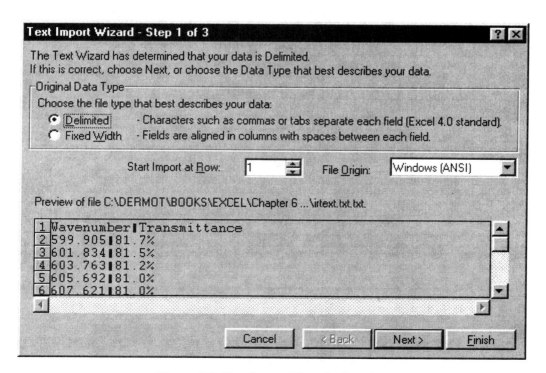

Figure 6.1. Text Import Wizard—Step 1.

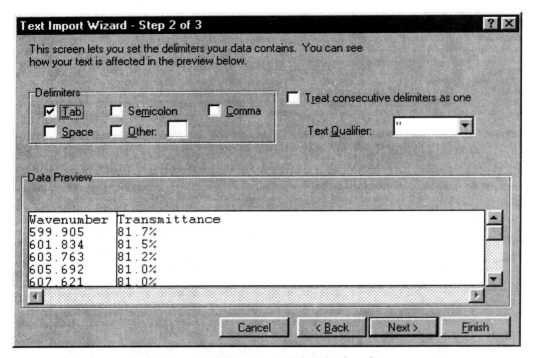

Figure 6.2. Text Import Wizard—Step 2.

Part of the resulting sheet is shown in Figure 6.4, with the wavenumber and transmittance data separated into two columns (A and B, respectively). The spectrum itself is shown in a graphical window in the same figure. It can be generated by following these steps.

- Select columns A and B (click on A and drag to B).
- With columns A and B highlighted, select the Chart Wizard button on the toolbar, and chart the data selecting the **Scattergraph with Line** option.
- Add labels and chart title as appropriate.
- When the Chart window appears, activate it by double clicking anywhere within it, select the data series by double clicking on the data, and switch off the data symbol so that only the line is plotted.
- Select the graph area by double clicking and select a white background and black line border.
- Double click on the x-axis and set the scale to the appropriate range by entering 600 in the Minimum field and 4000 in the Maximum field (see Fig. 6.5). Set the major unit at 500 and minor unit at 100. You must also select the

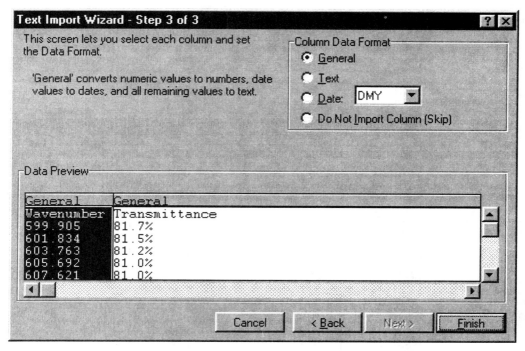

Figure 6.3. Text Import Wizard—Step 3.

Values in Reverse Order option as the default is to plot the data in increasing x-value order. For **y Axis Crosses at** enter 4000 (or select y Axis Crosses at Maximum Value). Finally, select the Patterns tab and set the tick mark type to "cross" (major) and "inside" (minor). The y-axis can be formatted in a similar manner.

It is relatively easy to manipulate the spectrum in order to further optimize the display, and this can be very beneficial for the preparation of figures for publications, 35-mm slides, reports, or *PowerPoint* presentations. For example, Figure 6.6 shows the same spectrum with the x-axis expanded to show the region from $600-1000$ cm^{-1} in more detail. This can be achieved simply by rescaling the axis display to the desired range, as described above. Once optimized, the spectrum can be copied to the Clipboard using Copy and Pasted into *PowerPoint* for slide generation, or into Word for integration into reports.

The flexibility of the Excel environment can also lead to other benefits over the sometimes rigid limitations of instrument software. For example, the Nicolet 205 FT-IR spectrometer comes with a built-in 3.5 in. floppy drive which is not DOS or Windows compatible, and the software restricts the number of spectra

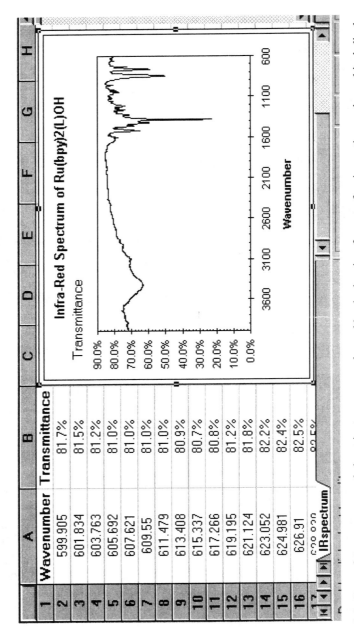

	A	B
1	**Wavenumber**	**Transmittance**
2	599.905	81.7%
3	601.834	81.5%
4	603.763	81.2%
5	605.692	81.0%
6	607.621	81.0%
7	609.55	81.0%
8	611.479	81.0%
9	613.408	80.9%
10	615.337	80.7%
11	617.266	80.8%
12	619.195	81.2%
13	621.124	81.8%
14	623.052	82.2%
15	624.981	82.4%
16	626.91	82.5%

Infra-Red Spectrum of Ru(bpy)2(L)OH

Figure 6.4. Part of *irspectrum* worksheet in *irspectrum.xls* workbook showing format after importing and graphical display of the spectrum.

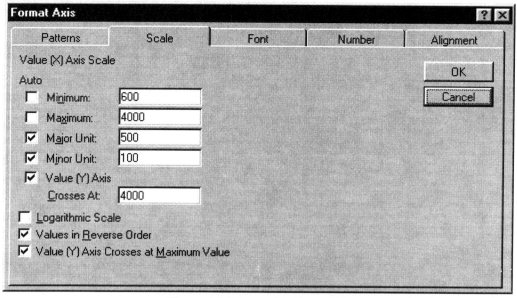

Figure 6.5. Formatting the x axis.

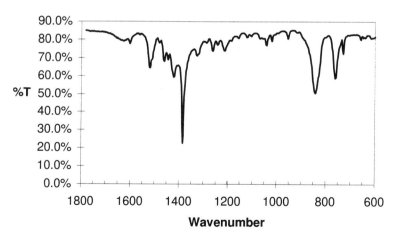

Figure 6.6. Expansion of the wavenumber scale of the IR spectrum to display fine features more clearly between 600 an d1800 cm^{-1}.

that can be superimposed to three, which is a severe limitation for kinetic and thermodynamic studies. Data transfer between the spectrometer and PC was accomplished via a serial port on the instrument using a standard file transfer protocol with the Windows Terminal application. Once available in a DOS compatible environment, spectra could be quickly displayed and multiple overlays generated (see Figure 6.7). In addition, the other Excel display modes, such as 3-D contour and 3-D Cartesian, could be employed or 2-D sections taken through the data (see Section 7.7).

6.4. SPECTRAL DATA PROCESSING

These days, data processing is becoming more and more complex, and teaching students the underlying principles can be difficult. Often students do not have a clear picture of what is really happening when they exercise various options. These include methods for improving the signal-to-noise (s/n) ratio or hidden feature extraction, such as digital filtering, wavelength averaging, internal referencing, differentiation, integration, spectral subtraction, digital baseline offsetting, and automatic identification of optimum wavelength range. Excel can provide an

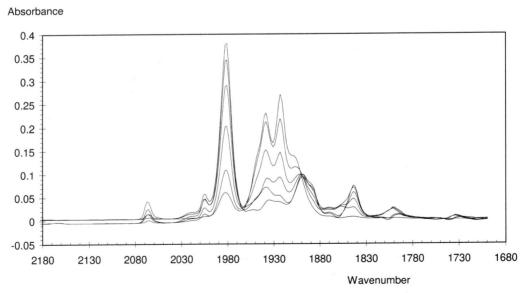

Figure 6.7. Six superimposed IR spectra obtained from a Nicolet 205 FT-IR spectrometer.

excellent environment for teaching these techniques or for implementing them where the instrument software is inadequate.

6.5. DIGITAL AVERAGING TECHNIQUES

Digital averaging techniques are now almost universally employed to improve the signal-to-noise ratio of instrumental data. They assume that the experimental data are equally spaced in terms of the x-axis (time, wavelength). The following gives some examples.

- *Ensemble Averaging.* This is the averaging of various numbers of complete scans, and it is normally restricted to instruments that produce complete data sets very quickly and reproducibly, such as photo-diode array UV-VIS spectrometers, FT-IR spectrometers, and NMR spectrometers. As this technique does not involve running a filtering window across a single data set, fine features will tend to be better resolved, rather than being merged into surrounding features.
- *Box-Car Averaging.* This is the averaging of n data points as they are acquired. It is useful in cases where the data acquisition has a large degree of redundancy (i.e., the data are being acquired much more quickly than is necessary to define the desired information content.
- *Moving Average Smooth.* In this technique, a filter window is moved along the data set after it has been acquired and stored in digital format. Although very simple to implement, it can be surprisingly effective; however, distortion of the signal can be severe.
- *Savitzky–Golay Smooth.* This algorithm is a least-squares polynomial fit that is applied to the data, using convolution weights to emphasize the importance of the central point in the filter window at the expense of the outer points, the logic being that the central point will more accurately reflect the true value of the signal at that point.

The worksheet *rawdata* in the workbook *fia.xls* contains part of a data set acquired from a flow-injection analysis system with an ion-selective electrode detector. Open this workbook up and graph out the data as described in previous sections using the Chart Wizard with Scattergraph option. Your result should look something like Figure 6.8. The data describe an FIA peak with a rather noisy baseline. Let us examine the effect of trying to reduce the baseline noise with the moving average and Savitzky–Golay digital filters.

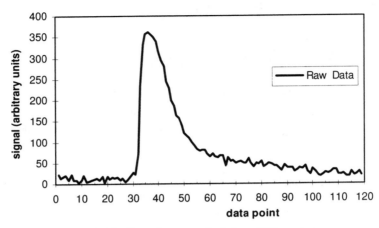

Figure 6.8. Raw data acquired from FIA system.

6.5.1. Moving Average Filter

In this filter, a window or bandwidth (n) of points is defined over which the data are to be averaged. The algorithm takes the mathematical form

$$\bar{x}_{\left(z + \frac{n-1}{2}\right)} = \sum_{i=z}^{i=z+n} \frac{x_i}{n} \tag{6.1}$$

where x_i is a particular point in the filter window and z identifies the first point in the window. The filter begins with z equal to the first point in the data set to be smoothed. The points from z to n are summed and the average defines the first value of the smoothed array, which begins after $(n-1)/2$ points from the beginning of the raw data set. The value of z is then incremented and the calculation repeated. In this manner, the window is moved along the raw data set, one point at a time, and the average of the n smoothed points stored as the midpoint of the new smoothed data array.

Unfortunately, this action results in $(n-1)/2$ points being "lost" from either end of the raw data set when smooth is completed. However, the unsmoothed original points can be plotted at each end in order to retain a full data set, although care should be exercised if further operations are to be performed on the data such as differentiation or alternative digital filters.

To implement the smooth, follow these steps.

• Open the workbook fia.xls and activate the sheet *rawdata*. The data start in cell A2 under the heading **Raw Data.** To implement a 15-point moving average smooth enter the formula **=SUM(A2:A16)/15** in cell B9. This places the average of cells A2 to A15 in B9.

- Select cell B9 and scroll to cell B113. Hold down the *SHFT* key and press the left mouse button to select the range B9 to B113 (the last raw data point is A120 and the filter will not operate over the final seven points).
- Use the **Edit+Fill_Down** command to fill the formula down over the selected range. The result is shown in Figure 6.9(a), column B. Note that the addresses of the cells in the formula from cell B9 are automatically incremented. The values returned by the filter are shown in Figure 6.9(b). Note that the raw data are copied across for the first and last seven points, but these can be omitted if desired.

Figure 6.9 also shows the formulas required to implement a seven-point moving average filter and the values returned, respectively, in column C.

To examine the effect of these filters on the signal, select the appropriate data range (e.g., C5 to C117 for the seven-point moving average) and plot as a scatter graph. The overall result is shown in Figure 6.10. Clearly, there is significant distortion of the analytical peak by the smoothing filter, which increases with increasing filter bandwidth. This is to be expected, as it will inevitably involve averaging part of the baseline with the rising portion of the peak and the peak maximum with the peak sides. The effect is seen more clearly in Figure 6.11, which can be easily generated by rescaling the x axis to the range 20 to 60. The filter severely distorts the peak shape by reducing and shifting the peak maximum and causing the peak rise to occur sooner. However, Figure 6.12 shows that the filter does improve the baseline noise. The performance of the filters in reducing baseline noise can be compared quantitatively through the standard deviations over a range where the baseline only is involved (e.g., points 8 to 24). These show a decrease from 5.106 for the raw data to 2.189 for the 7-point moving average smooth, to 0.825 for the 15-point moving average smooth.

This exercise clearly shows the dangers and benefits of smoothing data. The message is that filters should only be applied when necessary, and the overall effect should be carefully considered. In this case, the degree of distortion of the analytical peak is too severe to be justified by the improvement in the baseline signal. A better result would have been obtained if there had been more data defining the peak (e.g., the data had been acquired at a faster rate).

6.5.2. The Savitzky–Golay Filter

The Savitzky–Golay filter is a much gentler smooth than the moving average, as it involves applying a least-squares polynomial fit to the data. To implement a

	B	C	D
	15 pt MA	7 pt MA	7ptSG
1			
2	=A2	=A2	=A2
3	=A3	=A3	=A3
4	=A4	=A4	=A4
5	=A5	=SUM(A2:A8)/7	=(-A2*2+A3*3+A4*6+A5*7+A6*6+A7*3-A8*2)/21
6	=A6	=SUM(A3:A9)/7	=(-A3*2+A4*3+A5*6+A6*7+A7*6+A8*3-A9*2)/21
7	=A7	=SUM(A4:A10)/7	=(-A4*2+A5*3+A6*6+A7*7+A8*6+A9*3-A10*2)/21
8	=A8	=SUM(A5:A11)/7	=(-A5*2+A6*3+A7*6+A8*7+A9*6+A10*3-A11*2)/21
9	=SUM(A2:A16)/15	=SUM(A6:A12)/7	=(-A6*2+A7*3+A8*6+A9*7+A10*6+A11*3-A12*2)/21
10	=SUM(A3:A17)/15	=SUM(A7:A13)/7	=(-A7*2+A8*3+A9*6+A10*7+A11*6+A12*3-A13*2)/21
11	=SUM(A4:A18)/15	=SUM(A8:A14)/7	=(-A8*2+A9*3+A10*6+A11*7+A12*6+A13*3-A14*2)/21

(a)

	A	B	C	D
	Raw Data	15 pt MA	7 pt MA	7ptSG
1				
2	22	22	22	22
3	14	14	14	14
4	18	18	18	18
5	21	21	16.57143	16.80952
6	9	9	14.57143	17.19048
7	22	22	12.85714	15
8	10	10	11.71429	10.52381
9	8	12.6	11.57143	7.904762
10	2	11.13333	11	7.52381
11	10	10.2	8.857143	9.857143

(b)

Figure 6.9. Implementing the moving average and Savitzky–Golay filters. (a) The formulas used to implement the filters. (b) Values returned by 15-point and 7-point moving average filters, and a 7-point Savitzky–Golay filter with the initial points of the FIA Raw Data set.

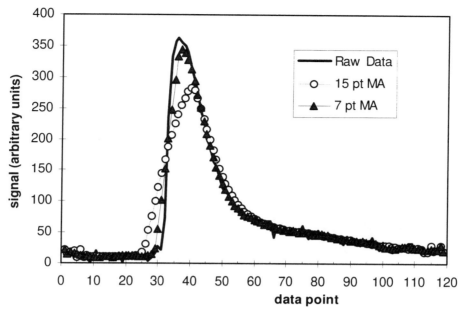

Figure 6.10. Overall effect of moving average filter on the data.

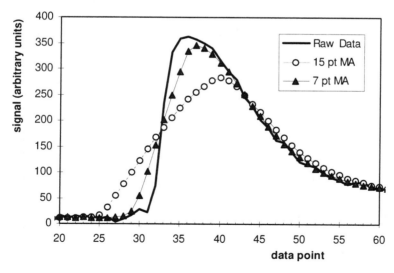

Figure 6.11. Effect of moving average filter on FIA peak.

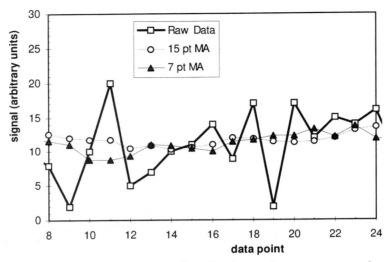

Figure 6.12. Improvement of baseline by moving average smooth.

seven-point Savitzky–Golay filter, a weighted average has to be calculated according to the Savitzky–Golay algorithm [2,3]. For a seven-point filter, the weighting coefficients are –2, 3, 6, 7, 6, 3, –2, respectively. Figure 6.9(a) shows the formula beginning in cell D5 and the values returned by the filter are shown in Figure 6.9(b). Figure 6.13 illustrates the effect of the seven-point Savitzky–Golay and seven-point moving average filters on the baseline. Clearly, the baseline noise

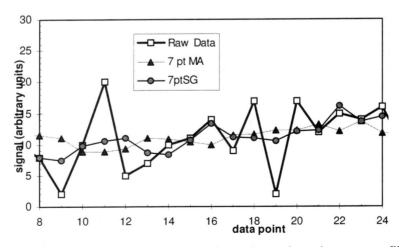

Figure 6.13. Relative Effect of 7-point Savitzky–Golay and moving average filters on FIA baseline.

amplitude is reduced by the Savitzky–Golay filter, but not as much as with the moving average (standard deviations over the range 8 to 24 are 2.189 and 1.403, respectively, compared to 5.106 for the raw data).

On the other hand, Figure 6.14 demonstrates that there is relatively little distortion of the analytical signal by the Savitzky–Golay filter in comparison to the moving average filter. The use of a seven-point Savitzky–Golay filter is consequently much easier to justify because the signal-to-noise ratio is improved by a factor of around 2.5 (ratio of the standard deviations) with almost no distortion of the analytical signal. Note that the complete filters and associated graphs can be found in the sheets *moving average* and *S–G filter* in the workbook *FIA.xls*.

6.5.3. Suggested Additional Activities

1. Given the array of raw data, students can implement to 7-point and 15-point moving average filters and compare the overall performance to that of a 7-point Savitzky–Golay filter.

2. A comprehensive set of weights for other Savitzky–Golay bandwidths and for differentiation of signals can be found in [1]. Students can apply other bandwidths and assess the impact in terms of peak distortion and improvement in signal-to-noise ratio.

3. Reference [1] also includes formulas for calculating the Savitzky–Golay

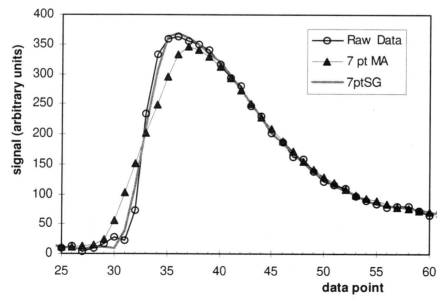

Figure 6.14. Effect of Savitzky–Golay and moving average filters on FIA peak.

weights for a particular bandwidth. These could be set up in an Excel sheet to develop a more flexible environment where the user could enter the bandwidth in a dialog box and the cell range of the raw data, and the filter is then automatically implemented with linked graphical display. Development of such an application would not be particularly difficult and would be a very useful tool for processing experimental data.

6.6. FEATURE REMOVAL THROUGH INTEGRATION

The moving average filter involves the summation and averaging of a bandwidth of data (the filter window) which is moved over the raw data set. In reality, this is a form of integration of the data. If the bandwidth is much larger than the x-axis unit base of a particular feature, then the feature may be completely removed by the filtering process. The following case study is designed to investigate this effect.

Case Study 6.1. Feature Removal by Filtering

- Open up the workbook *mafilter* and graph the data in column A. Your result should look like Figure 6.15.
- Examine the effect of using a 7-point, 15-point, 25-point, and 35-point moving average filter to smooth the data.
- Prepare a short report discussing which filter, if any, should be used. Pay particular attention to the number of points defining the peaks compared to the number of points in the filter window.

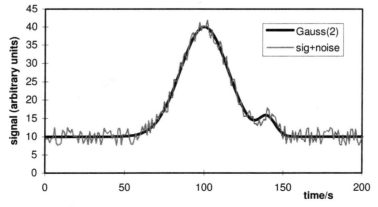

Figure 6.15. Simulation of two Gaussian features with noisy baseline.

Note: The Gaussian features in Figure 6.15 were generated in the sheet *gaussian and noise* in the workbook *mafilter.xls.* This sheet allows the user to vary the Gaussian parameters for the two features and the noise amplitude through the named variable *scale,* which is automatically added to the Gaussian features in the sheet and graphically displayed. The Gaussian parameters used in this case are given in Table 6.1. Peak 1 obviously refers to the major feature centered on 100 s, while peak 2 is the minor feature at 140 s. The resulting graphs should look like Figure 6.16. Complete worked examples with data and graphs can be found in the sheets *moving average* and *figures* in the workbook *mafilter.xls.*

This type of exercise is invaluable for reinforcing the importance of choosing the correct filter parameters for a smoothing operation, and showing how these digital filters actually work. In the next section, we shall look briefly at the opposite process, that of feature enhancement by differentiation.

6.7. FEATURE ENHANCEMENT THROUGH DIFFERENTIATION

In Section 6.6, we have seen how averaging techniques, which are essentially a form of integration, lead to the merging of features with a signal baseline, and, in certain cases, to complete removal of features. Differentiation has the opposite effect; it tends to amplify changes in the slope of a signal arising either from noise or from an analytical feature, which can then be used to enhance features in a signal and perhaps to extract hidden features. However, if differentiation is to be used successfully, the signal must be as noise free as possible, which in practice means that filtering and differentiation are often employed together, with filtering being used to smooth raw data prior to differentiation.

All data required for the following case study can be found in the workbook *feature extraction,* including Gaussian generation and addition of noise (sheets *test sets* and *more*) and the effect of differentiation and smoothing (sheet *dydx*).

Table 6.1. Gaussian Parameters Used to Generate Features in Figure 6.15.

Gaussian Peak/Parameter	1	2
Standard deviation/s	15	5
Height	30	5
Baseline offset	10	10
Time index of peak max/s	100	140

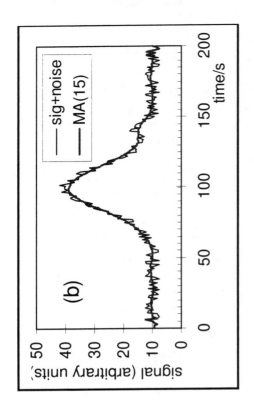

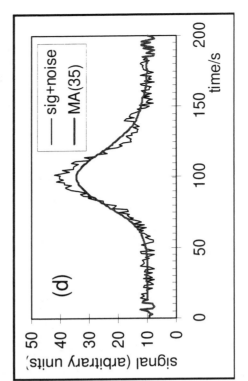

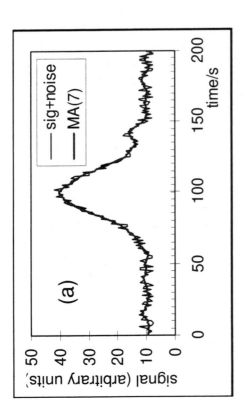

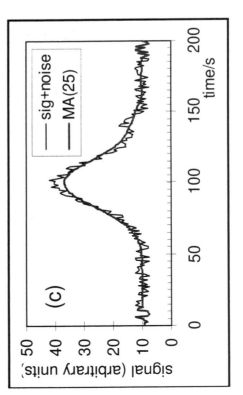

Figure 6.16. Feature removal by integration.

Case Study 6.2. Feature Enhancement and Removal Using Differentiation and Integration

- Open the workbook *feature extraction* and activate the sheet *rawdata.*
- Graph the **GAUSS(2)** and **SIG+NOISE** arrays (columns B and C) vs. **data index** (column A).
- Obtain the first differential of column C by simple differencing (e.g., place the formula **=C3-C2** in cell D3 and use **Edit+Fill_Down** to fill the formula down in column D over the range of the array of data in column C.
- Graph the data and interpret the results.

Notes: The signal in this case is formed from a dominant Gaussian peak centered at x value 100, with a second feature hidden within it centered at x value 140, shown in Figure 6.17. The hidden feature causes a slight asymmetry in the peak. The Gaussian parameters used to produce the waveform are listed in Table 6.2. Differentiation of the noisy signal produces a result as shown in Figure 6.18. Clearly there is no point in differentiating the noisy trace.

Case Study 6.3. Investigating the Effect of Moving Average Smooth on This Waveform

- Apply a moving average filter of bandwidth 15 to the data and summarize the result. You should produce a result similar to that shown in Figure 6.19. Why is the amount of distortion of the waveform much less than in the case of the FIA peak examined earlier?

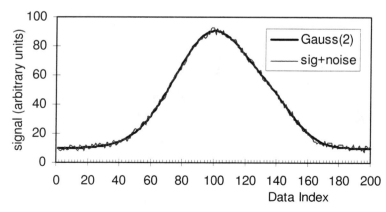

Figure 6.17. Gaussian signal before and after addition of simulated noise.

Table 6.2. Gaussian Parameters Used to Generate Data for Figure 6.17

Gaussian Peak/Parameter	1	2
Standard deviation/s	25	15
Height	80	15
Baseline offset	10	10
Time index of peak max/s	100	140

Case Study 6.4. Hidden Feature Extraction Using Differentiation

In this exercise we will use the first and second differential of the Gaussian signal, without added noise, to extract the hidden feature.

- Open the sheet *rawdata* in the workbook *feature extraction.*
- Do a scattergraph of column B [Gauss(2)] vs. Column A [Data Index]. You should end with the smooth waveform shown in Figure 6.20(a).
- Generate the first derivative of the waveform by simple differencing as described above (e.g., enter the formula **=B3-B2** in cell D3 and **Edit+Fill_Down** in column D to cover the range of data in column B. [*Note:* the full formula would normally be **=(B3-B2)/A3-A2),** as differentiation involves the difference in y as a function of the difference in x. However, as the difference in x (column A in this case) is unity, the formula simplifies to that shown above.]
- Obtain the second differential of the waveform by differentiating column D, for example, enter the formula **=D4-D3** in cell E4 and use

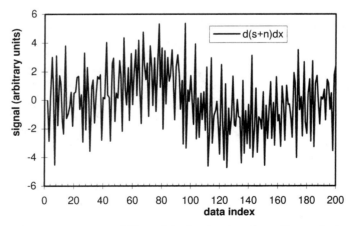

Figure 6.18. First differential of noisy data from Figure 6.17.

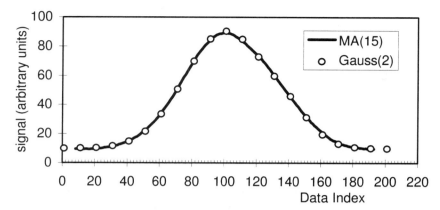

Figure 6.19. Effect of 15-point moving average smooth on the data.

Edit+Fill_Down to cover the range of data in column D. Enter appropriate headings for each column.

- Graph the data in columns D and E vs. column A. Your results should be similar to Figure 6.20(b) and 6.20(c), respectively.

Note: fully worked examples with data and graphs can be found in the sheet *dydx* in the workbook *feature extraction.*

This exercise, though simple, demonstrates the power of differentiation as a means of extracting information from signals about features that may be hidden in a more dominant feature. This technique is used routinely now to help interpretation of absorption bands in spectroscopy and peaks in chromatography that may contain hidden features. In Figure 6.20(a), there is just a hint that the peak is not symmetrical and may therefore contain a hidden feature. However, it is impossible to say what this feature is, where exactly it may be, and what its characteristics are. In contrast, Figure 6.20(b) shows the typical form expected of a Gaussian first differential, with the original Gaussian maximum being picked out at $x = 100$, where the first differential crosses zero. But clearly, there is something happening in the latter part of the differential, and the symmetrical waveform anticipated from the differential of a pure Gaussian is not obtained. This suggests there is some type of feature occurring between $x = 120$ and 150. This feature is even more clearly emphasized in the second differential [see Figure 6.20(c)]. The major feature is located by the minimum at $x = 100$, and the minor feature by the smaller minimum at $x = 140$, and its effect is more or less spread over a range of about plus or minus 15 x units. On consultation, we find that these data do reflect very well the parameters of the minor Gaussian hidden within the major feature (Table 6.2). This technique is

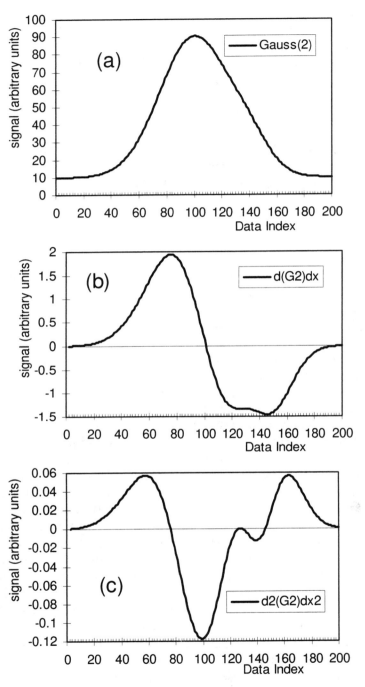

Figure 6.20. Hidden feature extraction by differentiation.

very useful for screening peaks in chromatography (peak purity detection) and absorbance bands in spectroscopy (e.g., location of hidden minor transitions in UV-VIS spectroscopy). Once located, a more rigorous modeling procedure could be used to obtain detailed information on the feature parameters (see Chapter 7 for more details on hidden Gaussian modeling using *Solver*).

In the next case study, we will use the second differential as a means of monitoring the extent of feature removal by varying filter bandwidths.

Case Study 6.5. Effect of Moving Average Bandwidth on Hidden Features

- Copy the data in columns A and B (x index and Gauss(2), respectively] and paste into a new sheet beginning at cell A2, and including the headings.
- In column C, obtain a smoothed version of the data in column B using a 15-point moving average smooth.
- Obtain the first and second differential of the smooth in columns D and E as described previously.
- Repeat steps 2 and 3, using a 25- and 35-point moving average, respectively.
- Graph each set of data vs. column A.

The overall result should be similar to that shown in Figure 6.21. Notice how, as the bandwidth of the filter increases, there is little apparent effect on the waveform itself [see (a), (d), and (g)]. In contrast, there are important changes in the differentials, and the features arising from the hidden Gaussian are almost completely erased, even in the second differential, by the use of a bandwidth of 35 [see Figure 6.21(i)].

This exercise once again emphasizes the power of even relatively simple digital signal processing tools to reinforce or to erase signal features, and therefore the potential for data enhancement or corruption. The lesson is that scientists should always retain the raw data that exist prior to using any digital signal processing, as well as the "optimized" final form which, if published, should list details of the signal processing tools used along with their parameters, so that the reader will be aware of this when interpreting published results.

6.8. CONCLUSION

The examples above show that Excel provides a very flexible platform for the display and processing of experimental data from analytical instruments. Once in Excel format, it is a relatively simple matter to cut and paste the graphics into

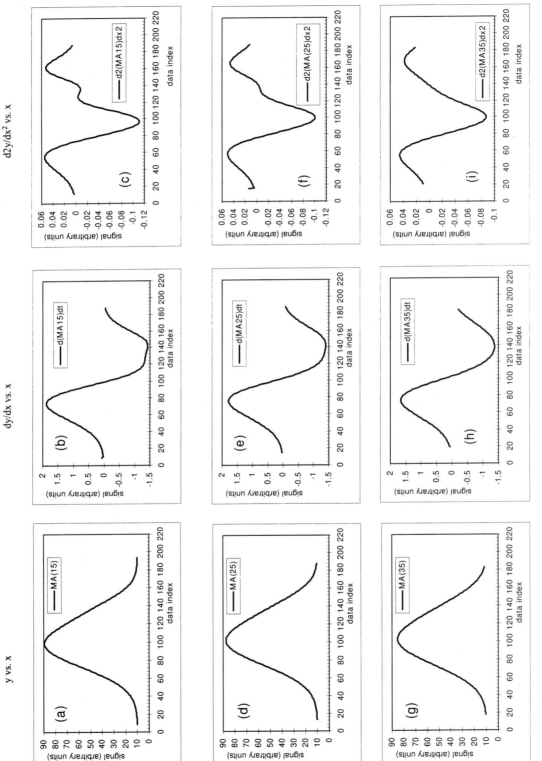

Figure 6.21. Effect of increasing moving average bandwidth on hidden feature.

Word for integration of documents and reports, and into *PowerPoint* for presentation graphics. Furthermore, linking the graphs during the transfer process automatically updates the graph in the linked package if the graph in *Excel* is changed in any way, which can be very useful if the raw data are being updated.

The processing of experimental data is described further in Chapter 7, which addresses the use of the optimization add-on *Solver* for performing least-squares nonlinear curve fitting on experimental data in order to extract values for experimental parameters.

REFERENCES

[1] Gorry, P. A. General Least-Squares Smoothing and Differentiation by the Convolution (Savitzky–Golay) Method. *Anal. Chem.* 62 (1990) 570–573.

[2] Savitzky, A. and Golay, M. J. E. Smoothing and Differentiation of Data by Simplified Least Squares Procedures. *Anal. Chem.* 36 (1964) 1627–1639.

[3] Steiner, J., Termonia, Y. and Deltour, J. Comments on Smoothing and Differentiation of Data by Simplified Least Square Procedure. *Anal. Chem.* 44 (1972) 1906–1909.

CHAPTER 7

PROCESSING EXPERIMENTAL DATA USING *SOLVER*

7.1. INTRODUCTION

The built-in regression tools available in Excel are useful for fitting relatively simple models (e.g., linear, exponential, polynomial) to experimental data. However, scientists may have particular models involving several parameters that they would like to fit to a data set. In fact, the case often arises where several models could be used, and the aim is to compare the fit, and then perhaps make theoretical interpretations based on the model. The built-in options offer very limited opportunities in this regard (except perhaps for linear regression), and when applied by students, can lead to very unsatisfactory fits to nonlinear data sets. In addition, these models only cover a few very simple options and give the user very limited control over the model parameters.

In contrast, *Solver* gives the user complete control over the parameter values and ranges over which they can be tested. The add-in module functions by comparing an array of data predicted by the model with an initial set of parameter values over a range of dependent variable values with a set of test or experimental values. A spreadsheet is designed to calculate the sum of squared residuals (SSR) between the two arrays [see equation (7.1)], and the parameter values are then varied according to an iterative search algorithm that seeks to minimize the error (SSR) between the two data sets. This means that the predicted data will eventually approximate the test or experimental data as closely as possible within the limits of the model and search algorithm used. In the following section, a series of case studies demonstrating the use of *Solver* are presented. These represent an excellent route to the introduction of curve fitting to undergraduate level students. The entire worksheets are laid out, with the test data and predicted data (using the model), a measure of the fit between the data sets in terms of the sum of squared residuals, and a graphical display that enables the user to follow the dynamics of the algorithm as it attempts to fit the predicted data set to the test data set. At regular intervals, the fitting process is interrupted by *Solver* and the user asked whether to proceed with further iterations. Press RETURN or click on OK to continue the process. Note how the value of the SSR decreases during the process until a limiting situation is reached at which the change in SSR or absolute value of SSR falls beneath a user-defined value. At this stage, the algorithm breaks out of the search process and presents the user with values for the model parameters that satisfy the thresholds set by the user. It should be noted that for real experimental data sets, which are always to a greater or lesser extent contaminated by noise, the value of the SSR will never be zero, but some value that ideally represents the effect of the noise on the signal. Of course, the user can define very complex polynomials or interpolations that will also model the noise very accurately, but these expressions are of little theoretical use if one is interested in the underlying signal. And in the end, it is once again up to the individual to judge whether a particular model is appropriate for fitting to a particular data set, and whether the fit obtained is acceptable or not.

Central to the successful application of *Solver* is the user's depth of understanding of the problem, since the user must arrange the raw data in a spreadsheet, enter the model and model parameters correctly, and finally initiate the model building process. Graphical windows enable the dynamics of the model building process to be observed, a feature that can be used to great effect in teaching the principles of curve fitting using iterative search algorithms.

7.2. USING *SOLVER*

The following steps describe the general principles of using *Solver* that are employed in the examples described later.

- The test or experimental data to be modeled (Y_i) must first be obtained. These are normally experimental results that may, or may not conform to a particular theoretical model. In the case of theoretical studies, the data are generated by means of the equation of interest. A suitable range of values can be quickly obtained by means of the Edit_Fill_Down command as described previously.
- A set of predicted values ($\hat{Y}_i$) is generated via the model.

- The goodness of fit between the test and predicted values is then determined by means of the sum of squared residuals (SSR), that is,

$$SSR = \sum (Y_i - \hat{Y}_i)^2 \tag{7.1}$$

- The values of the model parameters are then tested by the particular search algorithm selected within *Solver* to determine which changes generate a decrease in the SSR.
- The algorithm continues to search for variations in the model parameters that generate a decrease in SSR. The search process is terminated when one of several conditions is reached, such as time allocated, number of iterations, no further improvement can be obtained, or, ideally, the SSR falls below the acceptable threshold set by the user.

Figures 7.1 and 7.2 show the various procedures involved in finding the solution to a Gaussian peak generated using the well-known equation

$$y_i = H \exp\left[\frac{-(x_i - \bar{x})^2}{\sigma^2}\right] + B \tag{7.2}$$

where H = peak height above baseline,
  x_i = ith point on x axis,
  Y_i = value of the function at $x = x_i$,
  $\bar{x}$ = distance along x axis to peak maximum
  σ = standard deviation of the peak,
  B = baseline offset from zero.

Figure 7.1 shows the layout of a typical Excel spreadsheet prior to using *Solver* using data generated with equation (7.1). Column A contains the x-axis data, and column C contains the test set of Gaussian data generated via equation (7.1) using column A data and known values for the parameters H, $\bar{x}$, σ, and B, which have to be found by *Solver*. Column B contains data predicted using the Gaussian model [equation (7.2)] with the initial starting values of H, $\bar{x}$, σ, and B shown in cells G3 to G6 (note that $Z \equiv \bar{x}$ and $S \equiv \sigma$ on this sheet). For convenience, these parameters should be entered as named variables in Excel using the Insert_Name command.

In order to find the solution to the best-fit values of the Gaussian parameters, the user must present a target cell that contains the sum of squared residuals (SSR) between the test data (column C) and the predicted data (column B). The residuals are listed in column D and are obtained by subtracting equivalent cells in columns B and C. Column E contains the squared residuals, and cell G9 the

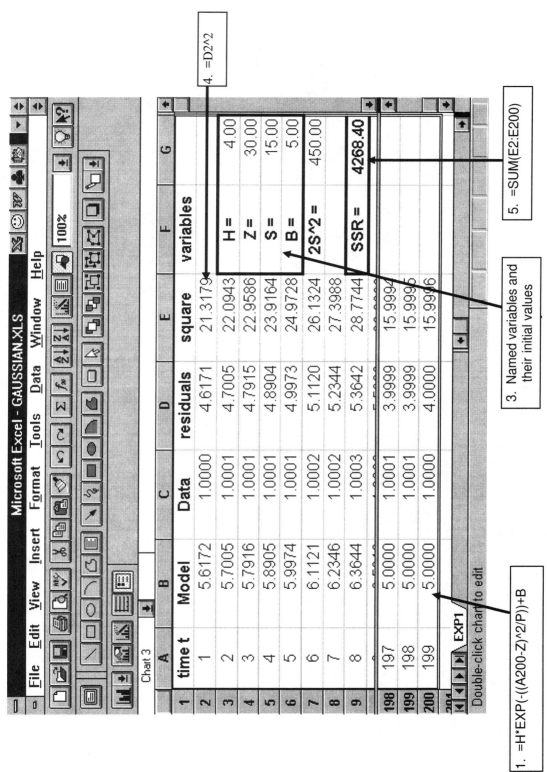

Figure 7.1. Typical layout of problem in Excel.

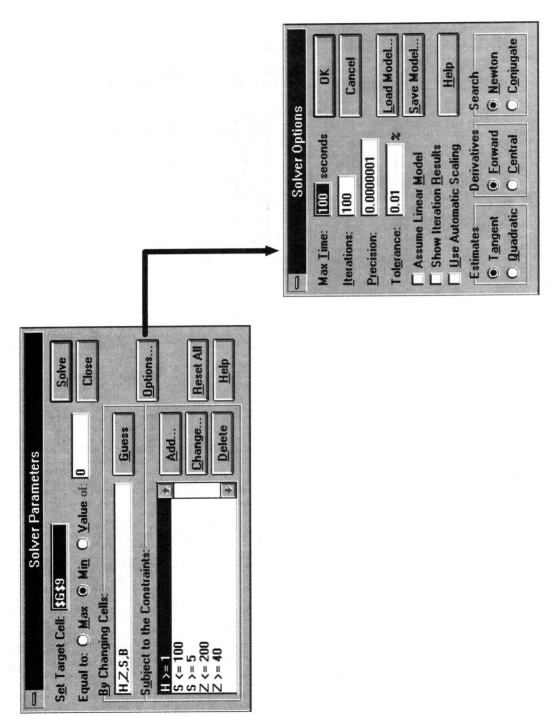

Figure 7.2. *Solver* starting conditions for single Gaussian peak fit problem.

sum of the squared residuals (i.e., the sum of cells E2 to E200). As this is a measure of the overall fit between test and model data sets, it is the quantity that will be minimized by *Solver.*

Solver is activated via the *Tools* menu. On activation, the user is presented with the panel shown in Figure 7.2, which in this case, is set up to solve the Gaussian problem described previously. The user options available are the following.

Set Target Cell. Specifies the target cell that can be maximized, minimized, or set to a certain value. In this case the target cell is G9 and it is set to be minimized.

By Changing Cells. Specifies the cells that will be varied by the search algorithm in order to minimize the number in cell G9, that is, cells G3 to G6 in this example.*

Subject to the Constraints. Constraints can be applied by the user to limit the search space explored by the optimization algorithm. As is usual in iterative search procedures based on gradient-type algorithms, efficiency is best if the search is initiated near the global solution of the problem and the unknown variables are restricted in value to realistic images. This strategy is required because these search algorithms may fail to locate the optimum solution to the problem, particularly if the error surface is rough (e.g., relatively large noise amplitude in the data), leading many local minima in which the algorithm may become trapped, or if the model is relatively complex and capable of returning several different combinations of parameter values that give good fits to the data. Clearly, therefore, one must be able to justify both the search strategy adopted and the parameter values returned by the model from a general knowledge of the problem under investigation and the quality of the experimental data.

Options. Displays the *Solver* Options dialog box (Figure 7.2) where the user can vary more features of the solution process. These are

Max Time. Limits the time allowed for the search process.

Iterations. Limits the number of iterations during the search process.

Precision. Sets the precision of the search process. This will determine the minimum change in the target cell that will cause the search process to stop. The best value for precision varies according to how far from the global optimum the initial search position is, and how smooth the error surface is.

Tolerance. Represents a percentage of error allowed in the optimal solution. A higher tolerance would tend to speed up the solution process but at the expense of accuracy in the final solution.

*In Figure 7.2, the named variables H, Z, S, and B, which are allocated to be the values in cells G3 to G6 are used in this field rather than the addresses.

Assume Linear Model. Speeds up the solution process but obviously should be used only if the relationship is linear.

Estimates. Specifies the approach used to obtain initial estimates of the variables. Briefly, these are

Tangent. Uses linear extrapolation from a tangent vector. That is, from a tangent, *Solver* extrapolates in different directions to identify which gives a minimum for the target cell. This identifies the next direction of the search process.

Quadratic. Uses quadratic (i.e., nonlinear) extrapolation, which can greatly improve results in very nonlinear problems at the expense of speed at arriving at an answer.

Derivatives. Forward and central differencing options exist for estimates of partial derivatives that give the gradient of the search at that point.

Search. Determines which search algorithm (Newton and conjugate) is used at each iteration. Both methods are dependent on the calculation of gradient values in the error surface at each stage in the iteration [1, 2]. The Newton method typically requires more memory than the conjugate search but requires fewer iterations. The conjugate search is useful if you have a complex problem (multiple model parameters, large data sets) and memory usage is a concern. We have found little difference in using either algorithm to date.

Load Model and Save Model. Used when more than one solver model is to be used with the worksheet.

Show Iteration Results. Gives the user a complete update of all data after each complete iteration. If the predicted and test data are graphed, the user can observe a dynamic display of the search process as it proceeds through each iteration. This is very useful in observing the progress of the algorithm, but can obviously greatly slow down the process where a large number of iterations are involved. If this option is not exercised, the search process proceeds automatically, reporting only the value of the target cell after each iteration.

Use Automatic Scaling. Useful in situations where data may be distorted through gross differences in the magnitude of various parameters in the search algorithm.

Figure 7.3 shows the results obtained with *Solver* for the Gaussian problem. The search begins [Figure 7.3(a)] with the predicted data (circles) being well displaced from the test Gaussian curve (solid line). Figure 7.3(b) shows the final positions of the test and predicted data, illustrating the excellent fit obtained. The values returned by *Solver* for the model parameters ($H = 10$, $\bar{x} = 20$, $\sigma = 100$, and $B = 1$) are equal to the values used to generate the test Gaussian curve. When the algorithm finds these values, the SSR drops to zero, indicating an exact fit be-

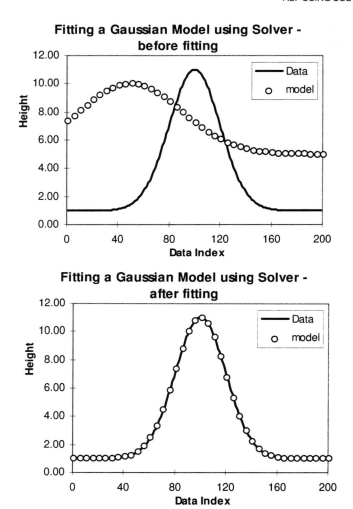

Figure 7.3. Fitting a Gaussian model using *Solver:* (a) before fitting; (b) after fitting.

tween the predicted and test data sets. This is not surprising as the test data set was constructed from the same model as the predicted data set. In order to generate a data set more like one that would be obtained during experimental work, the user could simulate the effect of noise by adding an array of numbers generated by the Excel random number generator to the test data set and repeat the modeling procedure (see workbook *gaussian and noise*). The worksheet *gaussian.xls* has the above problem preloaded, with the initial model parameters set well away from their true values as shown in Table 7.1.

Table 7.1. Initial and Final Values of Gaussian Parameters in Spreadsheet *Gaussian.xls*

Parameter	Initial Value	Final Value
H =	5.00	10.00
Z =	50.00	100.00
S =	40.00	20.00
B =	5.00	1.00

Note that in the spreadsheet and Table 7.1, **Z** represents the mean value of the Gaussian distribution (i.e., $\bar{x}$), and **S** represents the standard deviation (σ). The sheet has been set up with a macro activated by clicking the button **Start Solver** that will automatically run *Solver* with the preset model parameter values (Figure 7.2). Constraints have also been applied to the search in order to limit the error space to that defined by realistic parameter values. This considerably simplifies the search process and enables the algorithm to locate the true parameter values more quickly. It should be noted that the algorithm cannot locate the solution to the Gaussian test set from any starting set of parameter values. For example, the user should experiment by changing the initial parameter values so that there is no overlap between the test and predicted Gaussian distributions. In this situation, the algorithm may not be able to locate the test Gaussian if the variations in the parameter values during iterations do not generate a significant further decrease in the SSR. This will occur in cases where the search does not generate an overlap between the regions of the test and model Gaussian distributions, leading to no decrease in the SSR. If this happens over a number of iterations, the algorithm assumes that the best fit possible has been reached and the search process is terminated. The important lesson to the user is that even with a very simple fitting problem such as that described, the user must play a role in directing the search procedure through setting reasonable initial values for the model parameters and through constraints that limit the search to realistic ranges of values. In doing this, however, the user must always be able to justify the search strategy and demonstrate that it is not simply a reflection of personal prejudices, for example, limiting a search so that it cannot return a parameter value that the user does not wish it to have!

Note: The workbook *gaussian and noise* contains a Gaussian test set (representing a signal) with a noise contribution added (generated from the Excel random number generator). The noise amplitude can be varied by changing the value in cell J2 (labeled noisescale), simulating a variable signal-to-noise ratio, and the effect on the accuracy of the optimized parameters and the residual error of the fit can be investigated.

7.3. DECONVOLUTING TWO OVERLAPPING GAUSSIAN PEAKS

This is a situation that commonly occurs in chromatography and spectroscopy where peaks are commonly approximated by a Gaussian model (more of this later). Two species that almost coelute in chromatography will give rise to overlapping peaks, which may be difficult to detect if one of the peaks is significantly larger than the other. In this situation, the smaller peak may be manifested as a shoulder on the side of the larger peak. An analogous situation can occur in spectroscopy where two absorbance bands overlap.

Figure 7.4 illustrates such a situation that has been successfully modeled using *Solver*. In this example, both the test data and the model data were generated using equation (7.3), which is a simple summation of two Gaussian peaks and a baseline offset.

$$y_i = H_1 \exp\left[\frac{-(x_i - \bar{x}_1)^2}{2\sigma_1^2}\right] + H_2 \exp\left[\frac{-(x_i - \bar{x}_2)^2}{2\sigma_2^2}\right] + B \qquad (7.3)$$

where H_1 = height of peak 1 above baseline,
$\quad H_2$ = height of peak 2 above baseline,
$\quad \bar{x}_1$ = position of peak 1 maximum along x axis,
$\quad \bar{x}_2$ = position of peak 2 maximum along x axis,
$\quad \sigma_1$ = standard deviation of peak 1,
$\quad \sigma_2$ = standard deviation of peak 2,
$\quad B$ = baseline offset from zero on y axis,
$\quad x_i$ = ith point on x axis,
$\quad y_i$ = value of the function at $x = x_i$.

The problem is set up in the workbook *gaussian2.xls*. The test data set is generated on *sheet2* of this workbook and copied into column D in *sheet1*. This enables the user to vary the test set through changing the test set parameters on *sheet2,* for example, to investigate the limits over which the search algorithm can successfully deconvolute the two peaks. The effect is automatically transferred to the test set in *sheet1* and the search can once again be initiated through a macro that is activated by clicking on the **Start *Solver*** button. This option also allows the user to view the dynamics of the search process. However, as this is a relatively complex task with some seven parameters to be independently optimized, it will take quite a few iterations to solve the problem. To override the dynamic display, run *Solver* via the Tools command on the menu bar and deselect the option **Display Iteration Results** (see Figure 7.2). This allows *Solver* to proceed directly through iteration cycles, and although the user does not see the dynamics of the fitting process graphically, the value of the test cell (SSR) can be seen to decrease

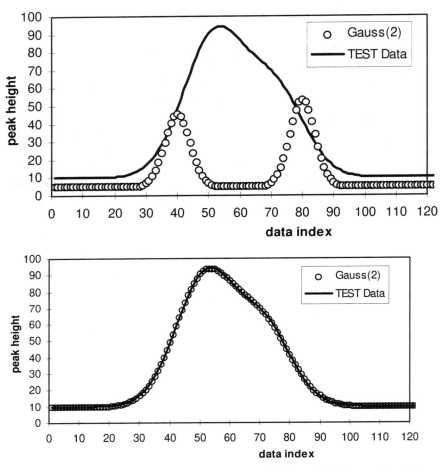

Figure 7.4. (a) Initial position of the test data set (solid line) and the predicted data set (open circles) prior to modeling with the double Gaussian expression. (b) Final positions of the test and predicted data sets showing successful modeling of the shoulder on the main peak.

steadily during optimization. Once again, the search returns an exact solution to the problems (SSR=0), and all parameters are successfully returned with the correct values ($H_1 = 80$, $\bar{x}_1 = 52$, $\sigma_1 = 10$, $H_2 = 45$, $\bar{x}_2 = 72$, $\sigma_2 = 9$, and $B = 5$).

Examples like these are very useful for teaching the principles of optimization and nonlinear least-squares fitting to graduate students, and in particular, in highlighting the need for direction by the user of the process because initial conditions can easily be set that do not allow the algorithm to locate the desired minimum.

7.4. FITTING EXPERIMENTAL DATA

The introduction deals with fitting test data that have been generated via a known equation, and the "true" values of the parameters required to be found by *Solver* are therefore known. This is a useful situation when one is exploring the limits of a modeling task. For example, one could vary the parameter values in the double Gaussian model to investigate when *Solver* is no longer capable of deconvoluting the data into two discrete Gaussian peaks. However, when dealing with experimental data, the situation is somewhat more complicated for several reasons.

1. There may be several possible theoretical models (or none at all) that can be applied to describe the data.
2. The data will have a certain amount of noise, which makes an exact fit impossible.
3. The optimum values can only be estimated rather than determined.

Hence, modeling experimental data sets is inherently more subjective and therefore more open to dispute than with synthetically generated data, which is what makes it a more interesting exercise.

7.5. CASE STUDIES INVOLVING EXPERIMENTAL DATA

7.5.1. Modeling Chromatography Peaks

Chromatography (and flow-injection) peaks are characterized by a Gaussian-type shape distorted by tailing that occurs on the falling portion of the peak. Models such as the exponentially modified Gaussian [3–6] and the tanks-in-series [7,8] have been developed in order to allow this distortion of the standard Gaussian peak shape to be described mathematically. This has important applications in the area of data storage (results can be described mathematically rather than stored as relatively large ASCII files), and in the analysis of peak purity (e.g., by comparing the shape parameters of experimental peaks to that of a typical peak obtained with the analyte under normal conditions). In this case study, we will examine the performance of three models that are commonly used to describe chromatography peaks, namely *Gaussian, Exponentially Modified Gaussian,* and *Tanks-in-Series*.

Gaussian Model. The HPLC data describing the peak can be found in the worksheet *gaussian.xls* (*sheet1*). The data were obtained from a UV-VIS detector as an ASCII file and imported into Excel. The peak, shown in Figure 7.5, begins at around 10 (x-axis scale is in seconds), rises quickly to maximum at around 16 s and decays to the original baseline by about 30 s. The trailing end of the peak is clearly visible between 20 and 30 s. The Gaussian model [equation (7.2)] is often used to approximate the shape of such peaks. *Sheet1* on the workbook *gaussian.xls* has the timescale of the experiment (one point per second) in column A, the HPLC experimental data (column C), and the results returned by the Gaussian model (column B), along with a graphical display of the peak. Part of *sheet2* of the workbook is shown in Table 7.2a, with column A containing the timescale of the peak (*x*-axis), column B showing the optimized results returned by the Gaussian model, and column C containing the HPLC experimental data. The residuals in column D were obtained by entering the formula =**B2−C2** into cell D2 and using the **Edit_Fill_Down** command to increment the formula down column D to cell D200 as described previously. In a similar manner, the formula =**D2^2** is entered into cell E2 and filled down to obtain the array of squared residuals. The sum of the array of squared residuals is obtained from cell G10 which contains the formula =**SUM(E2:E200).** This is the cell whose value is to be minimized by the search algorithm. Column H contains the error of the fit at each point on the peak expressed as a percent of the peak height. This is obtained by entering =**D2/1955*100,** where 1955 is the approximate peak height in bitnumbers. This normalizes the residual error in terms of the analytical signal, and hence enables a good feel for the size of the error to be obtained quickly.

The formulas for the first 10 cells in column B are shown in Table 7.2b. Enter the formula =**H*EXP(-((A2-Z)^2/P))+B** into cell B2 and fill down as before. Note that the variables in the formula **H, Z, B,** and **P** were defined previously as named variables with the addresses G3, G4, G6, and G7, respectively (use the Insert_Name_Define command and enter the name and address of each variable in turn), and that **P=2*S^2,** where **S** is the standard deviation of the Gaussian distribution (address G5).

Sheet3 on this workbook contains the percent error of all three methods to be investigated [see Figure 7.5(d)]. The error obtained with the Gaussian model varies between −8% to +6% of the peak height and oscillates sharply over the entire duration of the peak (ca. 10–30 s). Particularly worrying is the large and variable error around the position of the peak maximum (shown as a vertical black bar) which is the usual parameter for determining the amount of substance present. Clearly, the model only approximates the shape of the peak and returns an unacceptable error at the peak maximum.

Exponentially Modified Gaussian Model.

The exponentially modified Gaussian function (EMG) is the result of the convolution of a Gaussian function and an exponential decay. In Excel, this is achieved as follows; *Y(n)* represents an unconvoluted Gaussian data array of *n* points calculated according to equation (7.2). The difference *DY(m)* between a point *Y(m)* and the previous point *Y(m−1)* is easily obtained by subtracting the appropriate cells (note that the suffixes *(m)*

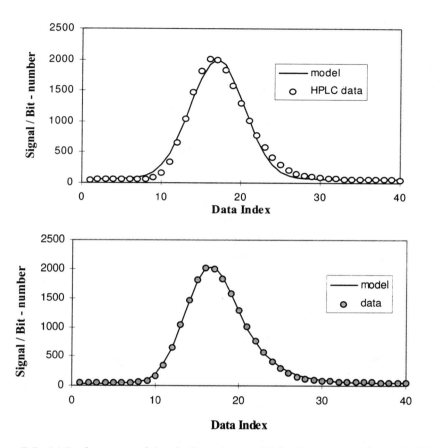

Figure 7.5. (a) Performance of simple Gaussian model for chromatography peak. (b) Performance of exponentially modified Gaussian model for chromatography peak.

and *(m–1)* denote the index or position of a particular point in the array). The convoluted array of points, EMG(m), is derived from the Y(m) array as follows:

$$EMG(m) = EMG(m-1) + \left[\frac{Y(m) - EMG(m-1)}{A} \right] \qquad (7.4)$$

where

$$A = \left[\frac{1}{1 - \exp(-W_2/\tau)} \right] W_1 \qquad (7.5)$$

τ = time constant of the exponential decay (where a time dependent function is being modeled), and W_1, W_2 = weighting factors, with W_2 normally set at unity.

On the spreadsheet, the Gaussian equation (7.2) is first used to create a Gauss-

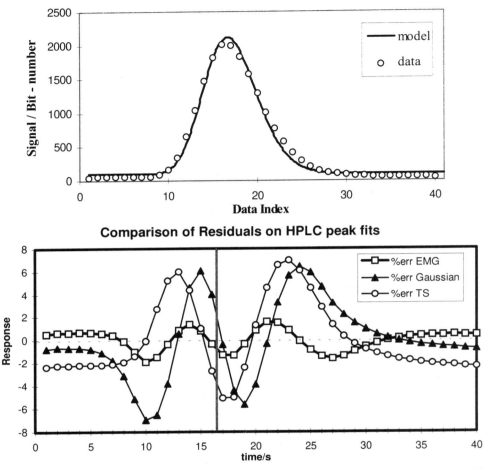

Figure 7.5. (c) Performance of Tanks-in-Series model for chromatography peak. (d) Comparison of residuals on HPLC peak fits for the three models.

ian data array over the range of interest. The first point on this array is then set equal to the first convoluted point *[EMG(1)]*. The second EMG point *[EMG(2)]* can then be calculated via equations (7.4) and (7.5). This procedure is repeated for the entire Gaussian array using the **Edit_Fill_Down** command in Excel.

The workbook *emghplc.xls* has the problem set out as shown in Table 7.3a. The Gaussian and Exponentially Modified Gaussian parameters listed in column I are defined as named variables with the values of the equivalent cells in column J. As with the Gaussian model, the parameter P is defined as $=2*S^{\wedge}2$, where S is the standard deviation of the Gaussian model. *WT1* is defined as cell K8, while *WT2* is ignored in this model. The formulas entered in the cells are shown in Table 7.3b and the parameter values in Table 7.3c.

Table 7.2a. Part of Sheet2 from the Workbook *Gaussian.xls*

	A	B	C	D	E	F	G	H	I
1	time t	model	HPLC Data	residuals	square	variables		%err Gaussian	
2	1	65.64	50.54	-15.1007	228.0302			-0.75315	
3	2	65.72	52.54	-13.1804	173.7218	H =	1940.506	-0.65737	
4	3	65.99	52.54	-13.4537	181.0031	Z =	16.97693	-0.67101	
5	4	66.85	53.54	-13.3096	177.1445	S =	3.382743	-0.66382	
6	5	69.29	53.54	-15.7524	248.1395	B =	65.61387	-0.78566	
7	6	75.65	53.54	-22.1047	488.6185	2s^2 =	22.88591	-1.10248	
8	7	90.68	55.54	-35.1346	1234.442			-1.75235	
9	8	122.98	61.54	-61.4435	3775.3			-3.06451	
10	9	185.96	85.54	-100.414	10082.98	SSR =	173549.6	-5.00818	
11	10	296.92	160.54	-136.378	18599.08			-6.80192	

Table 7.2b. Formulas for First 10 Cells in Column B

model
=H*EXP(-((A2-Z)^2/P))+B
=H*EXP(-((A3-Z)^2/P))+B
=H*EXP(-((A4-Z)^2/P))+B
=H*EXP(-((A5-Z)^2/P))+B
=H*EXP(-((A6-Z)^2/P))+B
=H*EXP(-((A7-Z)^2/P))+B
=H*EXP(-((A8-Z)^2/P))+B
=H*EXP(-((A9-Z)^2/P))+B
=H*EXP(-((A10-Z)^2/P))+B
=H*EXP(-((A11-Z)^2/P))+B

The optimized values of the parameters returned by *Solver* give a much improved fit with the experimental peak as shown by Figure 7.5(b), and the percent error in Figure 7.5(d). In the latter, the error is never greater than plus or minus 2%, and in the region of the peak maximum is never greater than plus or minus 1.5%, a significant improvement over the simple Gaussian model. A report summarizing the overall performance of *Solver* is shown in Table 7.4. *Solver* presents the user with this as an option on completion of the optimization task. It is automatically stored in a sheet called *answer report #* (where # = number of the report). This can be generated by the user if required and can be very useful in comparing the performance of different models or the quality of data sets. At a glance,

Table 7.3a. Layout of the Exponentially Modified Gaussian Model of HPLC Peak in Workbook *EMGHPLC.xls*, Sheet *EMG*

	A	B	C	D	E	F	G	H	I	J	K
1	t/s	y(t)	EMG(t)	Data	residua	res. sq.	%err EMG		**Gaussian Parameters**		
2	1	40.57	40.57	50.54	9.966	99.324	0.4971		S	2.571	2S^2 =P
3	2	40.58	40.58	52.54	11.964	143.147	0.5967		H	2709.307	13.222
4	3	40.62	40.59	52.54	11.952	142.845	0.5961		B	40.574	
5	4	40.85	40.66	53.54	12.880	165.887	0.6424		Z	15.020	
6	5	41.94	41.01	53.54	12.529	156.979	0.6249		**EMG Parameters**		
7	6	46.34	42.47	53.54	11.068	122.509	0.5520		dt	38.013	
8	7	61.48	47.69	55.54	7.854	61.679	0.3917		tau1	86.268	WT1
9	8	105.78	63.62	61.54	-2.083	4.340	-0.1039		A	3.645	1.127
10	9	215.39	105.26	85.54	-19.719	388.832	-0.9835				
11	10	443.48	198.04	160.54	-37.502	1406.411	-1.8704				
12	11	838.77	373.82	344.54	-29.275	857.046	-1.4601		ssr =	12163.05	

Table 7.3b. Formulas in Workbook *EMGHPLC.xls*, Sheet *EMG*

	A	B	C	D	E	F	G
1	t/s	y(t)	EMG(t)	Data	residuals	res. sq.	%err EMG
2	1	=H*EXP(-((A2-Z)^2/P))+B	=B2	50.54099	=D2-C2	=E2^2	=(E2/1955)*100
3	2	=H*EXP(-((A3-Z)^2/P))+B	=C2+(B3-C2)/A	52.54099	=D3-C3	=E3^2	=(E3/1955)*100
4	3	=H*EXP(-((A4-Z)^2/P))+B	=C3+(B4-C3)/A	52.54099	=D4-C4	=E4^2	=(E4/1955)*100
5	4	=H*EXP(-((A5-Z)^2/P))+B	=C4+(B5-C4)/A	53.54099	=D5-C5	=E5^2	=(E5/1955)*100
6	5	=H*EXP(-((A6-Z)^2/P))+B	=C5+(B6-C5)/A	53.54099	=D6-C6	=E6^2	=(E6/1955)*100
7	6	=H*EXP(-((A7-Z)^2/P))+B	=C6+(B7-C6)/A	53.54099	=D7-C7	=E7^2	=(E7/1955)*100
8	7	=H*EXP(-((A8-Z)^2/P))+B	=C7+(B8-C7)/A	55.54099	=D8-C8	=E8^2	=(E8/1955)*100
9	8	=H*EXP(-((A9-Z)^2/P))+B	=C8+(B9-C8)/A	61.54099	=D9-C9	=E9^2	=(E9/1955)*100
10	9	=H*EXP(-((A10-Z)^2/P))+B	=C9+(B10-C9)/A	85.54099	=D10-C10	=E10^2	=(E10/1955)*100

Table 7.3c. Values of Parameters in Workbook *EMGHPLC.xls*, Sheet *EMG*

I	J	K
Gaussian		
S	2.57119900982259	2S^2 =P
H	2709.30690856357	=2*J2^2
B	40.5738848393873	
Z	15.0197764878583	
EMG Para		
dt	38.0132980912358	
tau1	86.267564862422	WT1
A	=(1/(1-EXP(-dt/tau1)*WT1))	1.127475
ssr =	=SUM(F2:F61)	

one can see how the parameter values and the SSR have varied over the search. The workbook *emghplc.xls* contains the report shown in Table 7.4 (see sheet *answer report1*) and graphical displays of the fit to the experimental data on *sheet1*.

Tanks-in-Series Model. In this approach, the flow system is regarded as behaving as a series of mixing chambers that serve to distort the initial ideal square-wave concentration profile of the sample plug as it travels to the detector. The equation used in this instance is

$$f(t) = H\left[\left(\frac{1}{T_i(t/T_i)^{N-1}}\right)\left(\frac{1}{(N-1)!}\right)\exp(-t/T_i)\right] \tag{7.6}$$

where T_i = mean residence time of an element of fluid in any one mixing tank (i),
  N = number of tanks,
  t = time (x-axis index),
  H = scaling factor.

Table 7.4. Summary Answer Report Generated by *Solver* on Completion of the EMG Fit to the HPLC Data

Microsoft Excel 5.0 Answer Report
Worksheet: [CHROM3.XLS]EMG
Report Created: 29/7/94 20:36

Target Cell (Min)

Cell	Name	Original Value	Final Value
$I$8	ssr = Gaussian Parameters	497811.0497	12163.05363

Adjustable Cells

Cell	Name	Original Value	Final Value
$I$2	S	3.402766365	2.57119901
$I$3	H	1945.833065	2709.306909
$I$4	B	57.79239702	40.57388484
$I$5	Z	15.5	15.01977649
$N$2	dt	40	38.01329809
$N$3	tau1	60	86.26756486
$P$3	WT1	1	1.127475919

Constraints

Cell		Name	Cell Value	Formula	Status	Slack
$I$4	B		40.57388484	$I$4>=0	Not Binding	40.57388484

Table 7.5. Comparison of the SSRs Generated by Each of the Models of the HPLC Peak

Model	Gaussian	EMG	Tanks-in-Series
SSR	173549.6	12163.1	167309.3
Relative Performance	14.3	1	13.8

The workbook for this model is *tnkhplc.xls* and the performance is shown in Figure 7.5(c). Comparing the percent error generated by each model [Figure 7.5(d)], it is clear that the exponentially modified Gaussian model gives the best fit to the data, particularly in the region of the peak maximum. Table 7.5 compares the performance of each model over the entire data set in terms of the respective SSR values. Again, the EMG model is clearly the best, outperforming the Gaussian and Tanks-in-Series models by factors of 14.3 and 13.8, respectively. The EMG model can be interpreted in terms of the effect of diffusion and interaction of the sample plug with the tubing walls in a flow-injection system, which tends to distort the initial "square-wave" chemical impulse of the injection plug.

7.5.2. Modeling Fluorescence Decay Processes

One of the most common models used in processing experimental data is an exponential decay or growth. Examples include radioactive decay and reaction kinetics. In these situations, the objective is usually to study how the *rate* at which a reactant is disappearing (or product is appearing) varies with the amount or concentration of that substance. Very often, the kinetics will be first order (i.e., the rate is directly proportional to the amount of that substance present). For a substance A being converted to a product P,

$$A \rightarrow P \tag{7.7}$$

$$-\frac{d[A]}{dt} = k[A] \tag{7.8}$$

The negative sign on the rate of change of the concentration of A ($[A]$) with time (t) denotes that the amount of A present is decreasing with time. Experiments are directed at estimating the value of the rate constant (k). Linearization involves integrating the differential equation (7.8) which gives

$$\ln [A] = \ln [A]_0 - kt \tag{7.9}$$

where $[A]_0$ = the initial concentration of A. A plot of $\ln[A]$ vs. t will therefore be a straight line of slope $-k$, and intercept $\ln[A]_0$. Unfortunately, the equivalent strategy in *growth* processes involves estimation of the amount of a particular product at $t = \infty$, which requires waiting an infinite amount of time. Now for many processes, particularly fast reactions, $t \Rightarrow \infty$ in reasonable time periods, but this will not be the case for reactions that are slow, and for these cases, linearization will not be possible. An alternative approach is to fit a nonlinear model based on the first-order exponential expression to the data set. This is the approach adopted in the following examples.

The data set for the single exponential modeled in this case study was obtained from fluorescent emission decay lifetime measurements of the compound ruthenium-bis(2,2′-bipyridyl)(5-isothiocyanate-1,10-phenanthroline), that is, [Ru(bpy)$_2$ (NCSphen)]$^{+2}$ using a Q-switched ND-YAG laser. The samples were aerated and measurements were carried out in 0.1 M carbonate buffer (*p*H 9.6). The compound absorbs at 455 nm and is characterized by a fluorescence decay with a single time constant.

A double exponential model has been applied to fluorescent emission decay lifetime measurements of the same compound after attachment of bovine serum albumin via a thio-urea linkage [9]. Once again the samples were aerated and measurements carried out in 0.1 M carbonate buffer (*p*H 9.6). Binding of the protein has little effect on the excitation and absorbance wavelengths but lifetimes of fluorescent centers near the binding sites are affected, leading to differing sets of fluorescent emissions emanating from species bound in different environments. The models used in this investigation were general single and double exponential equations of the form

$$f(t) = [A(1 - e^{(-kt)}] + z \tag{7.10}$$

where A = pre-exponential factor,
$\quad\quad k$ = rate constant ($1/k$ = decay lifetime),
$\quad\quad t$ = time,
$\quad\quad z$ = baseline offset,

and for the double exponential,

$$f(t) = [A_1(1 - e^{(-k_1 t)}] + [A_2(1 - e^{(-k_2 t)}] + z \tag{7.11}$$

where A_1, A_2 = pre-exponential factors,
$\quad\quad k_1, k_2$ = rate constants,
$\quad\quad\quad t$ = time,
$\quad\quad\quad z$ = baseline offset.

Data obtained from the instrumentation are transformed to suit the above exponential models via in-house software prior to modeling, that is, the decaying signal obtained from the reducing concentration of the fluorescent reactant is inverted by the instrument to an increasing signal which reaches an exponentially limited maximum value of $A + z$ in the case of equation (7.10) and $A_1 + A_2 + z$ in the case of equation (7.11), for times approaching infinity. In fact the time constant of the fluorescence emission measured in these experiments is such that the process is more or less finished after about 10 μs.

The workbook *fluoexp1.xls* has been set up to model the first-order, single exponential data. The formulas used to model the single exponential data are shown in Table 7.6a and the resulting data in Table 7.6b. The *Solver* setup used to optimize the parameter values is given in Table 7.6c.

Table 7.6a.   Formulas Used to Model the Single Exponential Data[a]

	A	B	C	D	E	F	G	H	I
1	nde:	time	time/nS	data	model	residuals	res.sq.		variables
2	1	=A2*(0.000003/505)	=B2*10^9	3504	=a*(1-EXP(-k*B2))+z	=E2-D2	=F2^2		
3	2	=A3*(0.000003/505)	=B3*10^9	3880	=a*(1-EXP(-k*B3))+z	=E3-D3	=F3^2	a =	25934.45508641
4	3	=A4*(0.000003/505)	=B4*10^9	4016	=a*(1-EXP(-k*B4))+z	=E4-D4	=F4^2	k =	2221830.872648
5	4	=A5*(0.000003/505)	=B5*10^9	4376	=a*(1-EXP(-k*B5))+z	=E5-D5	=F5^2	z =	2677.183613698
6	5	=A6*(0.000003/505)	=B6*10^9	4608	=a*(1-EXP(-k*B6))+z	=E6-D6	=F6^2	1/k =	=1/k
7	6	=A7*(0.000003/505)	=B7*10^9	5016	=a*(1-EXP(-k*B7))+z	=E7-D7	=F7^2	ssr =	=SUM(G2:G300)

[a]Note that the timescale is calculated from the time taken to acquire a single point ($3 \times 10^{-6}/505$ seconds, columns A and B), and then converted into the nanosecond timescale (column C). The test or experimental data array is in column D and the model in column E. The residuals are calculated in column F and the squared residuals in column G. The parameters are in the box under columns H and I, and the SSR (to be minimized by *Solver*) is in cell I7. See Table 7.6c.

Table 7.6b.   Data from Single Exponential Model Returned after Optimization of the Fit

	A	B	C	D	E	F	G	H	I
1	index	time	time/nS	data	model	residuals	res.sq.	variables	
2	1	5.94E-09	5.94E+00	3504	3017.243	-486.757	236932.2		
3	2	1.19E-08	1.19E+01	3880	3352.844	-527.156	277893.6	a =	25934.46
4	3	1.78E-08	1.78E+01	4016	3684.044	-331.956	110194.8	k =	2221831
5	4	2.38E-08	2.38E+01	4376	4010.901	-365.099	133297	z =	2677.184
6	5	2.97E-08	2.97E+01	4608	4333.473	-274.527	75365.14	1/k =	4.5E-07
7	6	3.56E-08	3.56E+01	5016	4651.815	-364.185	132630.9	ssr =	3542181
8	7	4.16E-08	4.16E+01	5096	4965.982	-130.018	16904.56		

Table 7.6c. *Solver* **Set-up for Single Exponential Optimization**[a]

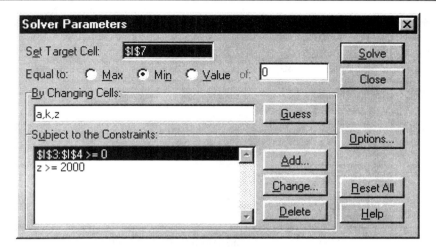

[a]Note that the target cell is the sum of squared residuals (cell I7—the dollar signs indicate that this is an *absolute* address). The cells to be changed are given by the named variables **a**, **k**, and **z**, set to the values of cells I3, I4, and I5, and which represent the pre-exponential factor, the rate constant, and the baseline offset in equation (7.10), respectively. Note also that constraints are applied to limit the search space by restricting **a** and **k** to positive values ($I$3:$I$4 >= 0) and restricting **z** to values greater than or equal to 2000, which is reasonable from the experimental data (see Figure 7.6a).

Figure 7.6(a) and 7.6(b) illustrate the fit obtained using the single exponential model [equation (7.6)] with the compound and the residuals of the fit, respectively. Note the time scale of the experiment (finished after 2 μs!). Clearly the model parameters returned by solver fit the data well, with the error (expressed as a percentage of the maximum response) never being greater than plus or minus 2%, which is quite acceptable given the noise on the original signal and the time base of the experiment. The time constant obtained from the fit ($\tau = 450$ ns) is typical of this material. Notice, too, how the error in the residuals decreases with time, a feature that arises from the relative difficulty in fitting the initial points of the exponential model (where the signal is changing most rapidly) compared to the later portion of the curve, where the signal is tending toward a constant value. Notice also how most of the variation in the residuals from about 1 μ onward follows an undulating pattern that is probably dominated by the noise component of the signal.

Figure 7.7 shows fits obtained with the protein-bound ruthenium compound using the single exponential model [equation (7.10)]. Initially, Figure 7.7(a) suggests quite a good fit to the data. However, comparing Figures 7.6(a) and 7.6(b) for the free (single environment) data, several differences are apparent.

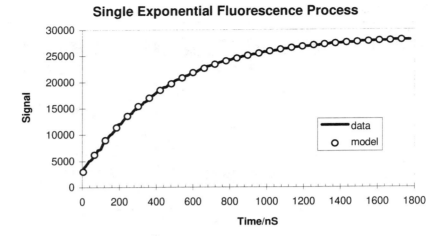

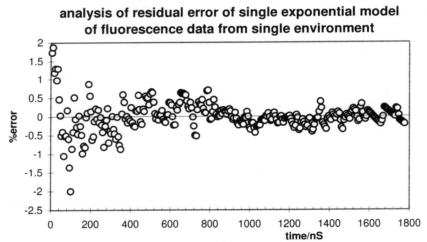

Figure 7.6. (a) Single exponential fluorescence process. (b) Analysis of residual error of single exponential model of fluorescence data from single environment.

1. The residual error is larger, particularly during the initial part of the curve, varying between abut plus or minus 5%, and the cumulative SSR is 4.44×10^7, compared to 3.54×10^6 for the free ligand.

2. There is clearly some underlying structure in the residual error in the range 0–1000 ns which suggests that the single exponential model is not describing the early form of the data accurately.

3. The residuals show a clear rising trend above about 1 µs, suggesting that longer lifetime processes are occurring that are not described by the model.

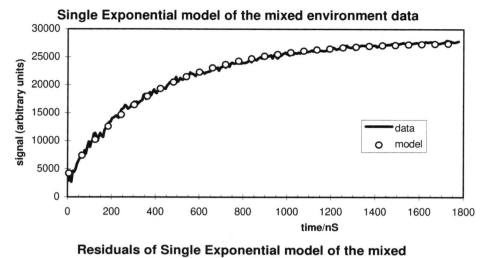

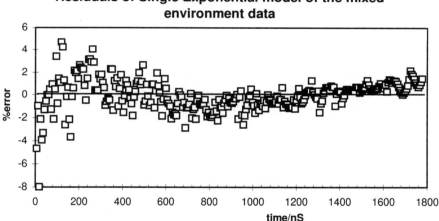

Figure 7.7. (a) Single exponential model of the mixed environment data. (b) Residuals of single exponential model of the mixed environment data.

4. The latter part of the residuals (above ca. 1 μs) show an underlying regular undulation [as does Figure 7.6(b)], which supports the suspicion that this feature arises from noise.

Fitting the double exponential model [Figure 7.7(c)] to the data improves matters somewhat. In particular, analysis of the residual error [Figure 7.7(d)] shows that the underlying structure in the early portion of the curve has been removed (the residuals are more symmetrically dispersed about 0% error) and the SSR is reduced to 3.54×10^7, yet the latter part of the curve still shows the underlying

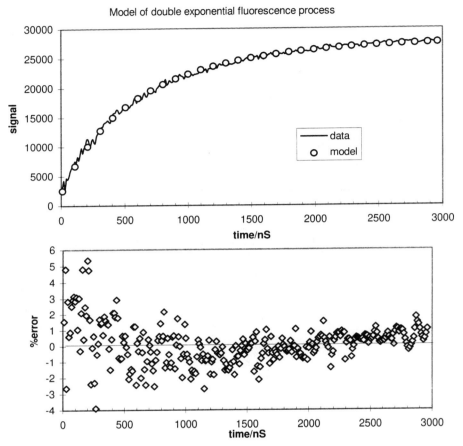

Figure 7.7. (c) Model of double exponential fluorescence process. (d) Error analysis of double exponential fit to fluorescence data from mixed environment.

rise and noise undulations. However, we should bear in mind that the experiment in Figure 7.6 does not run over the same time scale as that in Figure 7.7, and the underlying rise is most apparent in this later stage (this will also contribute to the larger SSR obtained with the models in Figure 7.7).

The double exponential model returns two time constants ($\tau_1 = 286$ ns, $\tau_2 = 833$ ns) with the error being within about plus or minus 3% outside the initial 200 ns of the curve. However, fitting two exponentials to data obtained in these experiments is almost certainly an approximation. Nonetheless, this type of approach can yield important information on the environment of the fluorescent center through examination on the relative populations of centers in the different environments, although this is a matter of considerable debate among specialists in the area. For example, in these experiments it is likely that the fluorescent centers exist in a number of different environments that modify the emission characteristics (e.g., the time constants). The fit with the double exponential model is reasonably good, and the time constants obtained are in the range expected. However, the curvature in the residuals suggests that there is some additional structure in the data. This in turn indicates that the fluorescent centers exist in two main environments, but this is probably not the complete picture. The message from this exercise is that curve fitting is almost more of an art than an exact science, and that careful examination and interpretation of the residuals is required, which can only be achieved with an in-depth knowledge of the theoretical background of the chemistry involved and the limits of the experimental method used to generate the data.

7.5.3. Modeling of Ion-Selective Electrode (ISE) Dynamic Response in Flow-Injection Analysis

We have also used *Solver* to look at the dynamic responses obtained with PVC membrane electrodes built into a flow cell for use with a flow-injection analysis system [10]. The peaks that are obtained from injections of the electrode detector on injection of various ion solutions can be captured digitally using an I/O card, and the data transferred into Excel as an ASCII file for display and further processing. One approach we have investigated is to use a logistic–sigmoid model to characterize the rising portion of the ISE response peak. This is achieved by extracting the data describing this portion of a peak in Excel and then using *Solver* to model the fit. The equation used is

$$E(t) = \left[\frac{a}{(1 + \exp[b(t-c)])^e} \right] + d \qquad (7.12)$$

where a = peak height/mV,
   b = slope coefficient,
   c = time from beginning of the peak to the inflection on the rise in seconds,
   d = baseline offset/mV,
   e = symmetry parameter for the sigmoid,
   $E(t)$ = electrode response at time t/mV,
   t = time in seconds.

This model can give some indication of the rate of ion uptake at the membrane surface as the sample plug passes and enable comparisons to be made for different experimental situations (varying concentration of the primary ion, effect of interferents, injection volume, flow rate, etc.). The model parameters in turn can be used as inputs in a further optimization of the instrumental operating conditions (e.g., optimize a combination of *a*, *b*, and *c* in terms of flow rate and injection volume). The workbooks *sigk@6r.xls* and *sigk@12r.xls* contain data from two peaks that have been analyzed using the sigmoid model. The data were obtained from a valinomycin-based PVC membrane that is known to be K^+ selective when two similar injections of a K^+ standard were made at two different flow rates (0.5 mL/min in *sigk@6r.xls* and 1.0 mL min in *sigk@12r.xls*. Table 7.7a shows the layout of the problem in the workbook *sigk@6r.xls* in terms of the various formulas used (a similar layout is used in *sigk@12r.xls*). As the experimental data were obtained at a rate of two points/s, the time scale can easily be calculated by dividing the data index by 2 (column B). The sigmoid model is coded into column C and the experimental data in column D. Column E calculates the residual error between the value predicted by the model and the experimental data, and this is squared in column F. The named variables used in the model equation (column C) are defined as **A, B, Z, D,** and **E** (cells G2 to G6), equivalent to *a, b, c, d,* and *e* in equation (7.12) and set to the values of cells H2 to H6, immediately to their right. The SSR is calculated in cell H7.

Table 7.7b shows how *Solver* is set up for this problem. The cell containing the SSR (H7) is set as the target cell to be minimized (once again, the dollar symbols indicate this an absolute address that will not be incremented during calculations). The cells to be acted upon by the algorithm are those containing the values of the named variables **A, B, Z, D,** and **E** (i.e., cells H2 to H6), and the only constraint is that **B** is less than or equal to zero. In this case, the values of the named variables were varied manually until a predicted curve similar to the experimental curve was obtained. *Solver* was then initiated with these as starting values and the optimized result obtained. This approach is useful as the algorithm is already fairly near the values that will generate the minimum error. However, the most satisfying (and scientifically robust) results are obtained when the same optimized values are obtained if the search is initiated from several random sets of initial conditions, with no constraints on the model parameters.

Table 7.7c shows the results obtained in *sigk@6r.xls* upon optimization. Note the relatively low SSR obtained, reflecting the good fit of the model. The same approach is used in workbook *sig@12k½r.xls,* the only difference in the two experiments being the flow rates (0.5 mL/min and 1.0 mL/min, respectively). Figure 7.8 shows the reasonably good fits obtained with the sigmoid model in both cases.

The model parameters returned by *Solver* for each flow rate are compared in Table 7.8. From these results, we can deduce that increasing the flow rate from 0.5 to 1.0 mL/min causes a slight reduction in the peak height (from 59.3 mV to 57.0 mV), an increase in the slope of the rise (given by the increased magnitude of *b*), a reduced time to the rise inflection (from 9.59 to 4.84 s), and a less sym-

Table 7.7a. Layout of Sigmoid Model in Workbook Sigk@6r.xls[a]

	A	B	C	D	E	F	G	H
1	Inde	Time/S	model	Data	residuals	SQ. Res.	Named	
2	1	=A2/2	=(A/(1+EXP(B*(A2-Z)))^E)+D	1.001	=D2-C2	=E2^2	A =	62.8064007742
3	2	=A3/2	=(A/(1+EXP(B*(A3-Z)))^E)+D	1.505	=D3-C3	=E3^2	B =	-0.10657421870
4	3	=A4/2	=(A/(1+EXP(B*(A4-Z)))^E)+D	2.236	=D4-C4	=E4^2	Z =	10.7771395772
5	4	=A5/2	=(A/(1+EXP(B*(A5-Z)))^E)+D	3.095	=D5-C5	=E5^2	D =	-4.35245187276
6	5	=A6/2	=(A/(1+EXP(B*(A6-Z)))^E)+D	4.189	=D6-C6	=E6^2	E =	1.89197721219
7	6	=A7/2	=(A/(1+EXP(B*(A7-Z)))^E)+D	5.297	=D7-C7	=E7^2	SSR =	=SUM(F2:F50)
8	7	=A8/2	=(A/(1+EXP(B*(A8-Z)))^E)+D	6.551	=D8-C8	=E8^2		

[a]The model is coded in column C.

Table 7.7b. *Solver* Settings for Sigk@6r.xls

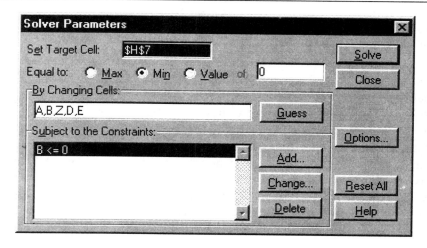

Table 7.7c. Results Returned by the Formulas Listed in Table 7.7b.

	A	B	C	D	E	F	G	H
1	Index	Time/S	model	Data	residuals	SQ. Res.	Named Variables	
2	1	0.5	0.585695708	1.001	0.415304292	0.172477655	A =	62.80640077
3	2	1	1.367385537	1.505	0.137614463	0.018937741	B =	-0.106574219
4	3	1.5	2.244066166	2.236	-0.008066166	6.5063E-05	Z =	10.77713958
5	4	2	3.220620265	3.095	-0.125620265	0.015780451	D =	-4.352451873
6	5	2.5	4.300744414	4.189	-0.111744414	0.012486814	E =	1.891977212
7	6	3	5.486672526	5.297	-0.189672526	0.035975667	SSR =	1.490259106
8	7	3.5	6.778925875	6.551	-0.227925875	0.051950204		

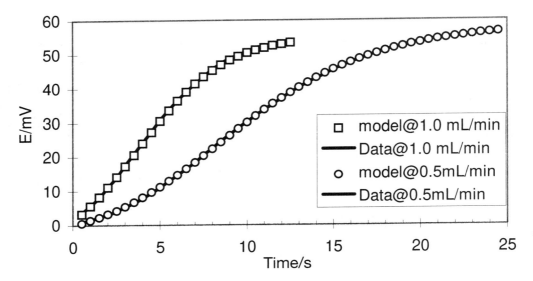

Figure 7.8. Fits obtained with a sigmoid model to rising portion of ISE peaks obtained at two flow rates.

metrical rise in terms of the sigmoid model (larger value for e). The results suggest that doubling the flow rate to 1.0 mL/min will not reduce the sensitivity of the response significantly (given by values for a) but will result in a faster response (the slope factor b is almost doubled and the time taken to get to inflection c almost halved). Characterizing peaks in this manner can be very useful for instrumental optimization purposes, as mentioned, and for describing peak shapes in terms of a few simple parameters. This characterization can be useful for processing large numbers of peaks and for identifying the possible presence of impurities through the definition of a "typical" analyte peak as possessing these parameters within certain limits.

Table 7.8. Model Parameters for Sigmoid Model of FIA Peak Rise[a]

	Flow Rate	
Model Parameters	1.0 mL/min	0.5 mL/min
a/mV	57.0	59.3
b	−0.178	−0.103
c/s	4.84	9.59
d/mV	−0.962	−1.14E
e	7.49	4.16

[a]Obtained with valinomycin ISE in an FIA system at two different flow rates (see Figure 7.8). Carrier composition: 5×10^{-2} M $MgCl_2$, 10^{-6} M KCl; sample composition: 5×10^{-2} M $MgCl_2$, 10^{-4} M KCl; injection volume: 150 μL.

7.5.4. Modeling the Nikolskii–Eisenman Equation

The effect of interferents on the response of an ion-selective electrode (ISE) is described by the semi-empirical Nikolskii–Eisenman equation

$$E = E^o + S \log(a_i + \sum K_{ij}^{pot} a_j^{z_i/z_j}) \tag{7.13}$$

where E is the measured response from the ISE-reference electrode cell, E^o is the standard cell potential, a_i, a_j, and z_i, z_j are the activity and charge of the primary ion (i) and interfering ions (j), respectively, S is the Nernst slope factor ($\approx 60/z_i$ mV per 10-fold change in a_i) and K_{ij}^{pot} is the potentiometric selectivity coefficient. The selectivity coefficient is an important parameter for describing the overall ability of the electrode to reject interfering ions in sample solutions, and for the electrode to function with acceptable error, a_i must dominate the summation with the parentheses in equation (7.13). Obviously this means that the selectivity coefficients should be very small for all possible interferents in order to drastically reduce their contribution to the overall signal.

Traditionally, selectivity coefficients are measured by the mixed and separate solution methods. The more realistic of these is the mixed solution method, in which the responses to solutions of varying primary ion activity but fixed interfering ion activity are measured.

Table 7.9 lists come cell potentials obtained with a calcium-selective electrode in pure calcium chloride solutions and in calcium chloride solutions with a fixed background activity of 0.1 M lithium chloride. Activity coefficients are calculated from the ionic strength and the Davies equation (see Chapter 5) and the activity of each cation obtained. The data are summarized in Table 7.9.

Figure 7.9 (bottom) shows a plot of the data, and the effect of the lithium on the electrode response can be clearly observed as a suppression of the response to calcium below about 10^{-3} mol dm^{-3} CaCl$_2$. Usually, the selectivity coefficient is

Table 7.9. Cell Potentials (mV) Obtained with a Ca Selective Electrode in Pure CaCl$_2$ Solutions [E(Ca)] and in CaCl$_2$ Solutions with a Fixed Background of 0.1 M LiCl [E(Ca+Li)]

$\log((a_{Ca})/\text{mol dm}^{-3})$	$a_{Ca}/\text{mol dm}^{-3}$	$E(Ca)/mV$	$E(Ca+Li)/mV$
−0.673	0.21232	30.10	30.1
−1.595	0.02541	4.45	05.6
−2.549	0.00283	−19.13	−16.7
−3.532	0.000294	−45.77	−34.3
−4.526	2.98E-05	−65.18	−41.6
−5.524	2.99E-06	−74.30	−40.4

estimated by extrapolating the horizontal portion of the mixed solution response until it intercepts with the extrapolated Nernstian portion of the response and by finding the intercept on the x axis from this point. However, this approach is very subjective, and the estimated coefficients are only rough guides to the performance of an electrode under real conditions.

A more satisfactory method is to fit the Nikolskii–Eisenman equation to the data for the mixed response. Figure 7.9 (top) shows part of the spreadsheet *selec-*

	A	B	C	D	E	F	G	H	I
1	**Mixed Solution Method**							**Named Variables**	
2	Calibration curve for tetraphosphine oxide calix[4]arene							S	25.6150
3	fixed background of 0.1M LiCl, variable CaCl2							Eo	46.9824
4	log(a_{Ca})	a_{Ca}	E(Ca)	E(Ca+Li)	model	res	sr	Kij	3.56E-0
5	-0.673	0.212324	30.1	30.1	29.76216	0.3378396	0.114136	Zi	
6	-1.595	0.02541	4.45	5.6	6.281365	-0.681365	0.464258	Zj	
7	-2.549	0.002825	-19.125	-16.7	-16.9889	0.2889062	0.083467	aj	0.
8	-3.532	0.000294	-45.7725	-34.3	-34.6542	0.3541712	0.125437		
9	-4.526	2.98E-05	-65.175	-41.6	-40.4505	-1.149528	1.321415		
10	-5.524	2.99E-06	-74.3	-40.4	-41.2506	0.8506177	0.72355		
11						ssr	2.832263		

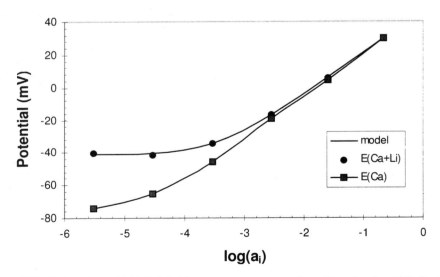

Figure 7.9. Estimation of Nikolskii–Eisenman parameters for a Ca selective ISE. Part of the workbook *selectivity* (top); fit (line) obtained to experimental data [•] (bottom).

tivity developed to do this. Cells C5:C10 and D5:D10 contain the potentials measured in the pure $CaCl_2$ and mixed $CaCl_2$/0.1 M KCl solutions, respectively. The model equation is entered in cells E5:E10 (i.e., **=E$_o$+S*LOG(B5+Kij*aj^(Zi/Zj))**) is entered in cell E5 and filled down). The residuals (res), squared residuals (sr) and sum of squared residuals (ssr) are calculated in the same way as in the previous examples, and the named variables are defined as cells H2:H7, and their values in cells I2:I7. *Solver* is set to minimize the sum of squared residuals (cell G11) by varying the value of the electrode slope (I2), standard cell potential (I3), and the selectivity coefficient (I4). The excellent fit obtained is shown in Figure 7.9 (bottom). This method returns values for the three model parameters (E^o = 46.98 mV, S = 25.62 mV/decade, K_{ij}^{pot} = 3.56 × 10^{-2}), which are in good agreement with the observed performance of the electrode. The sensitivity of the model to the selectivity coefficient value can be investigated by varying it by a certain amount and observing the effect on the fit. Figure 7.10 illustrates the sensitivity of the model to doubling and halving the selectivity coefficient value. Clearly, the model can define the selectivity coefficient to a much greater precision than this.

7.5.5. Determining Rate Constants for a Ligand Replacement Reaction

The following case study shows how experimental data can be imported into Excel, graphically displayed, and analyzed using *Solver* to obtain a first-order rate

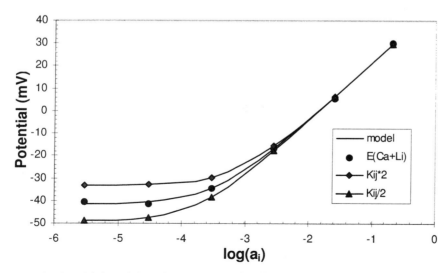

Figure 7.10. Sensitivity of the Nikolskii–Eisenman model to variations in the selectivity coefficient value.

constant. The data are UV-VIS scans obtained with a Hewlett-Packard HP 8452A photodiode array spectrophotometer, and the experiment consists of taking scans over the wavelength range 340–500 nm every 5 s during a photolysis experiment. The reaction involves replacement of a solvent molecule (e.g., ethanol, S) with a ligand (e.g., pyrimidine, py), which proceeds according to

$$Cr(CO)_5S + py \xrightarrow{h\nu} Cr(CO)_5py + S \qquad (7.14)$$

Using a large excess of ligand ensures that the reaction is pseudo-first order with respect to the Cr complex, and the rate constant can be estimated from the decrease in $[Cr(CO)_5S]$, or the increase in $[Cr(CO)_5py]$. The reactant has a strong absorbance with the maximum at about 450 nm, and during the experiment there is a large blue shift in the absorbance as the reaction proceeds, with an isosbestic point at about 430 nm (see Figure 7.11). The rate of the reaction is such that it can be followed by the photodiode array spectrometer. A plot of absorbance vs. time allows the observed rate constant to be determined using a first-order growth model;

$$absorbance = [A(1 - e^{-kt}] + B \qquad (7.15)$$

where A is a scaling factor, B is an offset, k is the rate constant and t is the time in seconds. Provided the Beer–Lambert law holds, absorbance will be directly proportional to concentration, and the rate constant can be obtained using equation (7.15). The raw data were imported as ASCII files into Excel and graphed using the scattergraph option to produce Figure 7.11. The data are available in the workbook *pdadata,* sheet *data.*

Data analysis was performed by extracting the absorbance vs. time data at 380 nm and pasting them to a new sheet (*model*). A plot of these data is shown in Figure 7.12 (top). The model was inserted in the form **=A*(1-EXP(-k*B1))+B,** where *A, B,* and *k* are as defined in equation (7.15), and **B1** is the time. This model is filled across the range required. *A, B,* and *k* are defined as named variables (for convenience), and initial guesses are made for their values (0.1 in each case). The residuals, squared residuals, and sum of squared residuals are calculated as described previously, and *Solver* activated to minimize the sum of squared residuals by varying *A, B,* and *k*. No constraints were applied to their values, and the default *Solver* settings were used. The solution [see Figure 7.12 (top)] is obtained within a few seconds, and the value of 0.01712 s^{-1} is in the expected range. Analysis of the residuals shows the error of the model to be reasonably small except in the initial part of the curve where the error is normally higher, as the measured quantity is changing most rapidly.

Furthermore, other wavelengths can be easily substituted in place of 380 nm

0 to 120 seconds (cycle time 5 seconds)

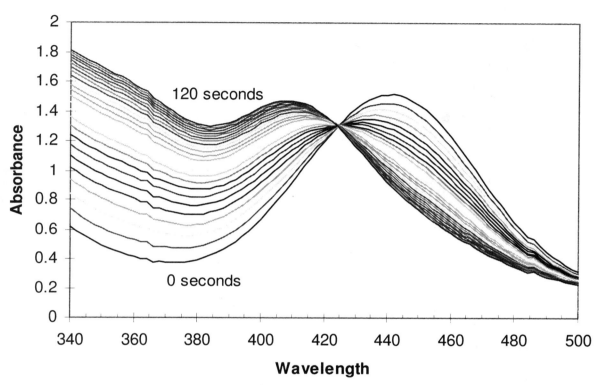

Figure 7.11. Successive UV-VIS spectra of the conversion of $[Cr(CO)_5S]$ to $[Cr(CO)_5py]$ (S = solvent molecule).

for the modeling procedure by pasting over the values in the sheet *model,* and the value of the rate constant checked. For example, Figure 7.13 shows the fit obtained using the data at 360 nm. Again, the fit is reasonably good, and the value of the rate constant returned by *Solver* is 0.01721 s⁻¹, which is close to the value obtained at 380 nm above, and is typical for this system at about 20°C. Alternatively, a first-order decay model could be used to obtain the rate constant from the data between 440 and 460 nm. This means that the rate constant can be easily determined over a number of wavelengths, rather than just at one, and the average reported along with the standard deviation. The optimum choice of waveband can be assessed by checking the standard deviation of the averaged rate constant.

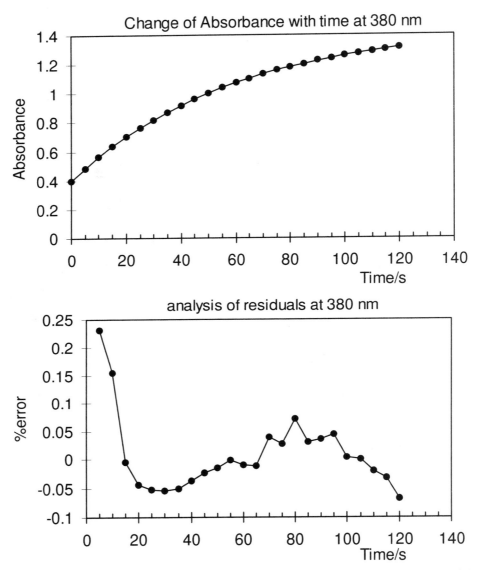

Figure 7.12. (top) Analysis of the change in absorbance at 380 nm as a function of time using a first order growth model [equation (7.15)] with experimental data [•], model as line, (bottom) analysis of residuals of the fit.

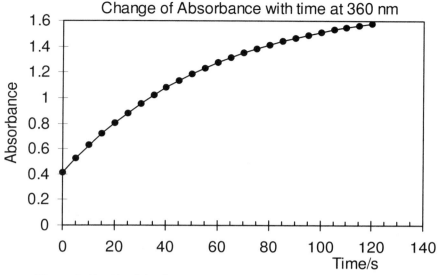

Figure 7.13. Fit of the first-order model to absorbance data at 360 nm.

7.5.6. Michaelis–Menten Enzyme Kinetics

Background. A common mechanism applied to the study of enzyme catalyses reactions is that proposed by Michaelis and Menten in 1913. Overall, the enzyme catalyzed reaction can be represented as

$$E + S \rightarrow P + E \tag{7.16}$$

where E is the enzyme, S is the substrate, and P is the product of the enzyme catalyzed reaction. However, many experiments show that the rate of product formation is dependent on the concentration of the enzyme, so, although the overall reaction is as shown in equation (7.16), there must be an addition stage that involves the enzyme. Michaelis and Menten proposed the following simple mechanism to explain the observed experimental dependency of rate on enzyme concentration;

$$E + S \underset{k_{-1}}{\overset{k_1}{\rightleftharpoons}} (ES) \overset{k_2}{\rightarrow} P + E \tag{7.17}$$

and the relationship between the reaction rate or velocity (V) and the enzyme and substrate concentrations is

$$V = \frac{V_{\max}[S]}{K_M + [S]} \tag{7.18}$$

where K_M is the Michaelis constant, and V_{max} is the maximum rate, which is in turn related to the total enzyme concentration (E_0) and the rate of decomposition of the bound enzyme–substrate intermediate (ES) by

$$V_{max} = k_2 E_0 \qquad (7.19)$$

Equation (7.18) can be linearized by taking reciprocals of each side and rearranging:

$$\frac{1}{V} = \left(\frac{K_M}{V_{max}}\right)\frac{1}{[S]} + \frac{1}{V_{max}} \qquad (7.20)$$

A plot of $1/V$ vs. $1/[S]$ (Lineweaver–Burk plot) should therefore give a straight line of slope K_M/V_{max} and intercept $1/V_{max}$. This has been the usual method for interpreting enzyme kinetics experimental data for many years. However, a double reciprocal plot like this tends to distort the data. For example, error bars are distorted, and data collected at equal substrate concentration intervals tend to bunch, producing a tendency to relatively large errors in regression data. In addition, as the reciprocal of the substrate concentration is used in the Lineweaver–Burke plot, a nonlinear variation in substrate concentration should be employed in the experiment to compensate for the bunching effect of the plot which obviously arises if a linear variation is used. Nonlinear modeling of the data is therefore an attractive alternative to the traditional Lineweaver–Burk plots and leads directly to simpler experimental designs.

Workbook Malate. The problem with the Lineweaver–Burk plot is well illustrated by the data set in the worksheet *linear-malate* [11], which can be found in the workbook *malate*. Figure 7.14 (top) shows the layout of the worksheet and the double reciprocal plot. The rate (V) and substrate concentration data are in columns A and B, respectively, and their reciprocals in columns C and D, respectively. The Lineweaver–Burk plot (bottom) shows the data to be very bunched near the x-axis, despite the attempt to space out the concentration intervals in a nonlinear manner. This is clearly very undesirable from the point of view of linear regression analysis. From the intercept and slope of the regression line, V_{max} and K_M are calculated in cells B12 and B15, and found to be 7.435 and 0.288, respectively.

The layout of a nonlinear model for a similar problem involving the enzyme creatinine kinase can be seen in Figure 7.15 (top). The substrate (creatinine) concentration [S] and rate V data are in cells A3:A8 and B3:B8, respectively, and the model entered into the adjacent cells in column C. For example, cell C3 contains the formula **=(Vmax*A3)/(Km+A3)**, with **Vmax** and **Km** defined as named

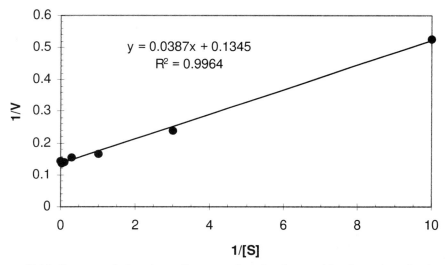

	A	B	C	D	
1	Malate Example from Noggle				
2	[S]	V	1/[S]	1/V	
3	0.1	1.9	10	0.526316	
4	0.333	4.2	3.003003	0.238095	
5	1	6.1	1	0.163934	
6	3.33	6.5	0.3003	0.153846	
7	10	7.2	0.1	0.138889	
8	33.3	7.4	0.03003	0.135135	
9	100	6.9	0.01	0.144928	
10					
11	Intercept=	0.1345			
12	Vmax=	7.434944			
13					
14	Slope=	0.0387			
15	Km=	0.287732			

$y = 0.0387x + 0.1345$
$R^2 = 0.9964$

Figure 7.14. Layout of the sheet *linear malate* in the workbook *malate* (top) and Lineweaver–Burk plot (bottom).

variables in cells H4 and H5 (see worksheet *nonlinear creatinine* in the workbook *malate*). The residuals are calculated in cells D3:D8, and the squared residuals in cells E3:E8. Finally, the sum of squared residuals is calculated in cell E9 using the formula **=SUM(E3:E8).** The nonlinear plot [Figure 7.15 (bottom)] is obtained by plotting the experimental points using a scatterplot without joining the points.

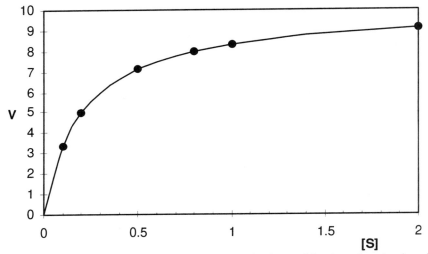

	A	B	C	D	E	F	G	H
1	Creatinine Kinase							
2	[S]	V	Model	residuals	SR			
3	0.1	3.33	3.331642	-0.00164	2.7E-06		Named Variables	
4	0.2	5	4.997991	0.002009	4.03E-06		Vmax	9.999151
5	0.5	7.14	7.140958	-0.00096	9.17E-07		Km	0.200127
6	0.8	8	7.998307	0.001693	2.87E-06			
7	1	8.33	8.331746	-0.00175	3.05E-06			
8	2	9.09	9.089614	0.000386	1.49E-07			
9				SSR=	1.37E-05			
10								

Figure 7.15. Layout of the sheet *nonlinear creatine* in the workbook *malate* (top) and the fit to the nonlinear plot (bottom).

The model is the solid line, which is generated by calculating the value of the model over the range of substrate concentrations of interest (0–2 mmol/L) using a larger number of points (20 in this case) and graphing as a scattergraph without data symbols but with points joined with a smoothed line. These points are calculated in cells A12:B32. *Solver* is initiated with cell E9 as the target cell whose value is to be minimized, by varying cells H4 and H5 (named variables **Vmax** and **Km**). No constraints need be set, and the values of **Vmax** and **Km** can be varied manually until the model line is in the approximate region of the data. Selecting *show iteration results* enables the user to observe the dynamics of the fitting process. The final result is as shown in Figure 7.15 (bottom). An excellent fit to

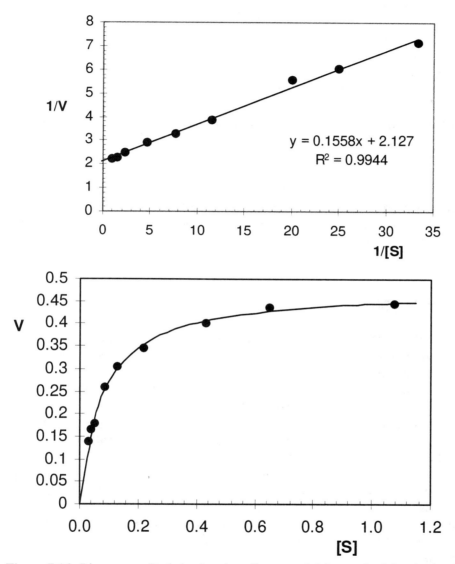

Figure 7.16. Lineweaver–Burk (top) and nonlinear model (bottom) of data in the sheets *linear amylase* and *nonlinear amylase* on the workbook *amylase.*

the data is obtained and the values for V_{max} and K_M are found to be 9.9992 and 0.2001, respectively, which is in excellent agreement with results obtained using the Lineweaver–Burk approach (10.0020 and 0.2003, respectively).*

Clearly, it is relatively easy to obtain high quality data from enzyme kinetics experiments without having to resort to distorting the data by taking reciprocals. This is also apparent from the following example.

Workbook Amylase. Lineweaver–Burk and nonlinear modeling analysis are compared in the workbook *amylase* for data obtained with the enzyme amylase [11]. Figure 7.16 shows the Lineweaver–Burk plot (top), and the nonlinear model (bottom). Once again, the values returned by both methods are similar at 0.470 and 0.477 for V_{max}, and 0.073 and 0.076 for K_M (nonlinear data second) despite the use of substrate concentration intervals which were designed with the Lineweaver–Burk method in mind. For a very useful discussion on the errors involved in fitting enzyme kinetics data, and the propagation of errors in curve fitting generally, see [11].

7.6. CONCLUSION

The examples in this chapter demonstrate that Excel can be used for quite advanced curve fitting and data analysis. The message however, is that you must have a considerable knowledge of the subject of the study if the results of the analysis are to be interpreted properly. The graphical display of the fitting process that can be activated from the **Solver_Options** menu (show iteration results) enables the user to follow the dynamics of the algorithm as it attempts to minimize the contents of the target cell (the residual error between the test and predicted data). The test sets for each case presented in this chapter are available on diskette and can be readily adapted for teaching purposes with differing levels of complexity. For example, students can be given just the test data and the model equation, and asked to analyze the data using *Solver.* Alternatively, students can be given the complete workbook and asked to explore the limits of the algorithm by setting various initial starting values for the model parameters or setting different constraints.

Solver is undoubtedly a useful data analysis tool, and while it does not compare in performance with dedicated chemometrics or advanced scientific data analysis packages, it does allow the user to model quite complex data sets with relative freedom. The fact that it becomes bundled with Excel (most users are probably unaware of its existence!) makes it a very attractive route to teaching the

---

*Lineweaver–Burk analysis of the creatinine kinase data can be found in the worksheet *linear creatinine* in the workbook *malate.*

basic principles of curve fitting and data analysis. In addition, we can assume that the current trend of making Excel more appealing to scientific users will continue, and, as the package is unlikely to be discontinued and has a huge user base (unlike some of the more dedicated packages), we can assume that time invested in learning to use *Solver* will not be wasted.

REFERENCES

[1] Hostetter, G. H., Santina, M. S., and D'Carpio-Montalvo, P. *Analytical, Numerical and Computational Methods for Science and Engineering.* Prentice-Hall, Englewood Cliffs, NJ 1991.

[2] Press, W. H., Flannery, B. D., Teukolsky, S. A., and Vetterling, W. T. *Numerical Recipes.* Cambridge University Press, 1990.

[3] Berthod, A. Mathematical series for signal modeling using exponentially modified functions. *Anal. Chem.* 63 (1991) 1879.

[4] Delley, R. Series for the exponentially modified Gaussian peak shape. *Anal. Chem.* 57 (1985) 388.

[5] Foley, J. P., Dorsey, J. G. Equations for calculation of chromatographic figures of merit for ideal and skewed peaks. *Anal. Chem.* 55 (1983) 730.

[6] Grushka, E. Characterization of exponentially modified peaks in chromatography. *Anal. Chem.* 44 (1972) 1733.

[7] Ruzicka, J. and Hansen, E. H. *Flow Injection Analysis.* Wiley, New York, 1988.

[8] Lee, O., Dumont, G. A., Tournier, P. and Wade, A. P. *Anal. Chem.* 66 (1994) 971.

[9] Ryan, E., O'Kennedy, R., Feeny, M., Kelly, J. and Vos, J. G. Covalent Linkage of Ruthenium Polypyridyl Compounds to Poly-L-Lysine Albumins and Immunoglobulin G. *Bioconjugate Chem.* 3(4) (1992) 285.

[10] Walsh, S. and Diamond, D. Non-linear curve fitting using Microsoft Excel Solver. *Talanta* 42 (1995) 561.

[11] Noggle, J. H. *Practical Curve Fitting and Data Analysis.* Prentice Hall 1993.

CHAPTER 8

CASE STUDIES INVOLVING VBA

The elements of VBA allow the development of user interfaces to spreadsheets for the entry of data and echoing results. This is done through the use of the dialog sheet that includes the dialog box. When a dialog sheet is active, its specific toolbar, the Forms toolbar, is displayed with all the available objects. The dialog box is a container for the objects of VBA that are used to direct and reflect entry of cells of the spreadsheet. Some examples of these objects are list boxes, drop-down boxes, option buttons, and check boxes. It is important to remember that VBA works on objects; they are controlled with VBA and used to accomplish tasks. The following case studies demonstrate the valuable features of VBA in designing spreadsheets.

8.1. TITRATION OF A STRONG ACID WITH A STRONG BASE

This case study involves the principles of pH calculation and determination of a titration curve. Constructing the curves can be laborious and the spreadsheet offers an ideal vehicle of simplifying this process. Comparing concentrations of acid, titrant, and initial acid aliquot volume to the constructed curve is advantageous to understanding the fundamental principles. The Excel dialog sheet and VBA can be used to achieve a user-controllable "front end" to the calculations. The user can then easily change the initial parameters of the titration and see immediate response in the titration curve.

Case Study 8.1. Visual Examination of the Titration Curve of a Strong Acid

The acid is titrated with a strong base by adding a VBA dialog sheet for user interaction with the titration parameters.

Setting up the Spreadsheet

- Open a new workbook and Save As *Chap8.xls*.
- Rename the first sheet tab to *data*.
- In A1 enter the title **Strong acid Titrated with Strong Base.**
- Type in **[HCl]=** and **[NaOH]=** in cells A3:A4.
- Enter in C3: **Vol of HCl.**
- Define names for these three variables. Select B3 and choose Insert_Name_Define. The Refers to: input box should display **Data!$B$3** and the Names contain **HCl.** Accept these defaults and choose Add. For the NaOH variable, either edit the Names input box with **NaOH** and Refers to with **Data!$B$4** or leave the Define Name dialog box, select B4, and rechoose Insert_Name_Define. The default input entries should be as above. For Vol of HCl, edit Names to **Vol_A** and reference the cell to **Data!$D$3.** Remember, defined names cannot contain spaces. As the formulas to follow will be entered based on these defined names, enter sample data in these cells so when the formulas are entered they do not display error messages; for example, enter in 0.1, 0.1, and 25 respectively for **[HCl], [NaOH],** and **Vol_A.** Later we shall return to these cells to link the user selections for the titration parameters from the dialog box to the calculation section of the spreadsheet.
- Enter data column headings in A6:D6: **Vol of Titrant, Moles Titrant, Moles HCl, pH, [H+].** Adjust the column widths accordingly. Select these entries, format **B** (bold), and place a heading border (R3C1 option of the Borders drop-down menu) on them.
- Use Edit_Fill_Series to enter the Vol of Titrant (mL) data from **1** to **55**, incrementing by **1.**
- Enter the formula for Moles Titrant (moles) in B7 using the defined name for [NaOH]: **=NaOH*A7/1000.** Split the window and Fill_Down the formula.
- Enter the formula for Moles HCl (moles) in C7 using the defined name for [HCl]: **=(HCl*Vol_A/1000)-B7.** Split the window and Fill_Down the formula.

The calculation of pH (starting with D7) will depend on whether the number of moles of HCl are negative, zero, or positive. In each case, a different calculation is required. If the moles of HCl are positive, the pH is determined from [H$^+$] but if they are negative, the pH is determined by difference from the moles of [OH$^-$] using K_w. If the moles of HCl are zero, the *p*H is neutral. Therefore an **If** statement is used to distinguish the three cases. This is best constructed through the Function Wizard where inputting the arguments of the If function are easily controlled. The If will be nested with two further Ifs for the three cases and these will be listed in numerical order.

- Select D7, then the Function Wizard and choose the **If** function. The cursor should be in the Logical_Test input box. The first logical test is if the number of moles of HCl are less than zero. This translates to **C7<0**. Tab to the Value_If_True box and enter the expression of the *p*H: **14+LOG(C7/-(A7/1000+Vol_A/1000))**. The false condition applies both to when the number of moles of HCl are equal to zero and greater than zero. Hence the false argument is another If or nested If.
- Tab to the Value_If_False box and select the Function Wizard icon to the left of this input box. Choose the **If** function. Here the Logical_Test is **C7=0**. Value_If_True is **7,** a neutral solution. The false argument of this If dialog box is the second nested If.
- Select the Function Wizard icon in the Value_If_False box and choose the **If** function. The final Logical_Test is when the moles of HCl are greater than zero, **C7>0**. The formula for *p*H if this is true is **-(LOG(C7/(A7/1000+Vol_A/1000)))**. There is no further condition, so the Value_If_False can be left blank, though best practice is to put in an error message entry. Enter **error!**. The complete formula for the entry in D7 is

  =IF(C7<0,14+LOG(C7/-(A7/1000+Vol_A/1000)),
  IF(C7=0,7,IF(C7>0,-(LOG(C7/(A7/1000+Vol_A/1000))),"error!")))

- Select OK in each of the two nested If dialog boxes and choose Finish in the last.
- Split the window, select the range for the formula, and Fill_Down. The spreadsheet is shown in Figure 8.1.

Graphing

- Select the **Vol of Titrant** and **pH** data using the CTRL key to aid selecting nonadjacent cells. Draw an XY Scattergraph of format 6 using the Chart Wizard. Ensure the correct entries have been made in Step 4. Label the graph accordingly in the last step.

Figure 8.2. shows a sample graph for this spreadsheet.

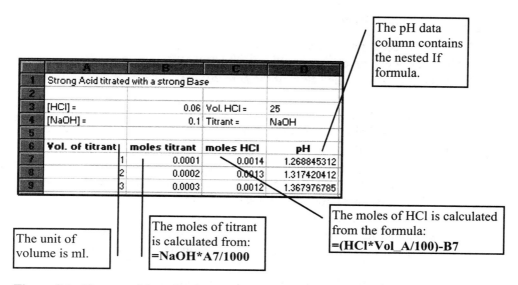

Figure 8.1. The spreadsheet for the calculation of *p*H from the defined titration parameters of cells B3, B4, and D3. When these parameters are changed, the data in the columns is recalculated.

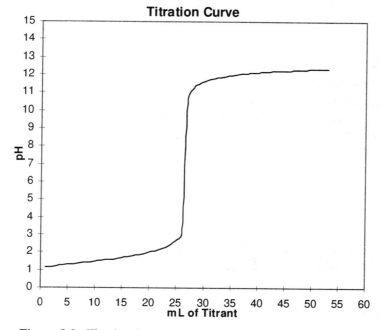

Figure 8.2. The titration curve for strong acid with strong base.

The Excel Dialog Sheet

The dialog sheet is introduced to allow the user to change the parameters of the titration and view the effects on the graph. VBA objects will be introduced directing control to specified cells of the spreadsheet.

- Right click the next sheet tab and insert a dialog sheet. By default, this sheet has a dialog box with OK and CANCEL buttons. The Forms toolbar (Figure 8.3) is displayed when a dialog sheet is active.
- Move the mouse pointer over the tools of the Forms toolbar to acquaint yourself with the types of objects that can be placed on a dialog sheet. Find the drop-down tool ▦ and select it. Draw a box on the dialog sheet. Use this tool to draw two more drop-down boxes and align the three above one another as shown in Figure 8.4. Each of these drop-down boxes is to correspond to each parameter of the titration to be altered: molarities of HCl and NaOH, and the aliquot volume. In a moment, the choices to be selected from the drop-down boxes will be assigned to it.
- Use the Label tool ▦ to place a label above each drop-down box, entering text for each of the three variables to be changed in the drop-down box.
- Select the Group Box tool and draw a box around the drop-downs and labels. Use Figure 8.3 to find this tool.

Assigning Values to the Dialog Sheet Objects

Each drop-down object of the dialog sheet needs to have values assigned; these will become the choices when the dialog sheet is run. This is accomplished by the preparation of a VLOOKUP table on the spreadsheet, containing the ranges for the molarities and volumes along with an indexing number. The index number is required because the choices of the drop-down are registered through

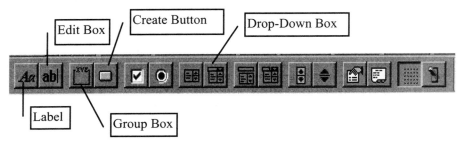

Figure 8.3. The Forms toolbar for dialog sheet development.

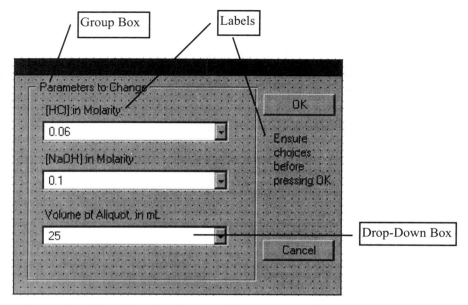

Figure 8.4. The dialog box for controlling the entry of the titration parameters.

this number, not the actual choice selected. A VLOOKUP table is necessary to associate the index number back to the choice selected. These values are then used in the formulas of the spreadsheet, as they have already been defined as names.

- Select the data sheet and scroll down to the end of the data, about row 65.
- Enter the titles **# from dialog, [HCl], [NaOH],** and **Vol of Aliquot,** beginning in A65.
- Enter the data according to Figure 8.5. This is the data that will appear when the drop-down boxes of the dialog sheet are selected. The concentrations of the acid and base will range from 0.01 M to 0.35 M and the aliquot volume from 5 to 50 mL. These three ranges will be assigned to the drop-down object of the dialog sheet.
- The table to the right of the VLOOKUP tab shows the cells (G66:G68) of the spreadsheet where the choices of the drop-down boxes are linked. These values are placed on the spreadsheet when the dialog box is run. Enter labels to the left to identify these cells as indicated in Figure 8.5. All the data is in place to link to the drop-down objects of the dialog sheets.
- Activate the Dialog Sheet tab, Dialog1, and select the drop-down object

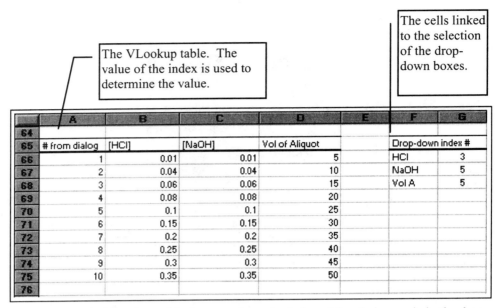

Figure 8.5. The VLookup table that associates the index of the choice made in the drop-down of the dialog sheet to the value. To the right is the table of the cells linked to the selection of the drop-down. This is used by the VLOOKUP function to associate the index back to the selected value.

for the [HCl]. Use the keypress CTRL + 1 to bring up the Format Object dialog box. Select the Control tab. The entries for controlling the drop-down are shown in Figure 8.6. The drop-down box gets its information from the spreadsheet through Input Range for displaying choices and outputs the choice made by the user back to the spreadsheet through Cell Link. Type in or select the range of concentrations from the VLOOKUP table on the data sheet for entry in the Input Range box, B66:B75.

- Tab to Cell Link and enter the cell reference of the cell where the output of the dialog box is to be stored, G66. This output is an index or number corresponding to the choice made by the user from the displayed list. This index will be converted back to the actual value through use of the VLOOKUP table.
- The Drop Down Lines option sets the number of choices to be displayed from the drop-down when it is selected. If this number is less than the total choices in the list, the user will have to scroll to view all the choices. If it is more than or equal to the list range, all the choices will be visible, with a few blank lines in the latter case. The list range in this data sheet is

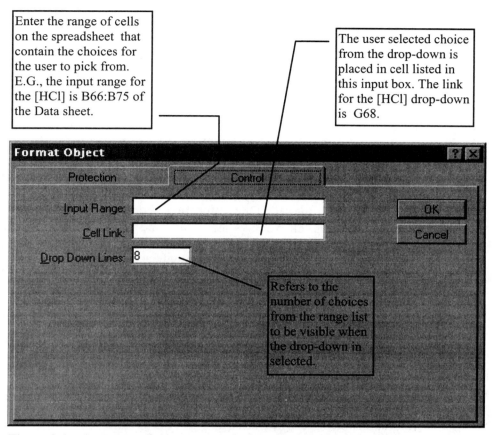

Enter the range of cells on the spreadsheet that contain the choices for the user to pick from. E.G., the input range for the [HCl] is B66:B75 of the Data sheet.

The user selected choice from the drop-down is placed in cell listed in this input box. The link for the [HCl] drop-down is G68.

Refers to the number of choices from the range list to be visible when the drop-down in selected.

Figure 8.6. The Format dialog box outlining the entries for controlling input to and output from the drop-down object.

10; choose a number less than this so as not to crowd the dialog box when displayed. Press OK when all entries are complete.

- Select the remaining two drop-down objects and repeat the described steps to format them by entering the appropriate cell references from the spreadsheet.

Associating Drop-Down Choices with Spreadsheet Cells

The titration parameter choices made by the user need to be entered into the calculation part of the spreadsheet. At the moment these choices are as indices in G66:G68 and are blank because the dialog sheet has not been run yet. The VLOOKUP table is already prepared for their conversion to titration parameter values. Now we just need to enter the VLOOKUP function for the titration parameters in those cells where their names have been defined. These are for **[HCl]** in B3, **[NaOH]** in B4, and **Vol of Titrant** in D3.

- Starting with [HCl], select B3, then the Function Wizard. Choose the VLOOKUP function (Figure 8.7) under the Lookup & Reference category. The lookup_value for [HCl] is G66, the cell referenced in the cell link input box of the HCl concentration drop-down box. Enter the range of the VLOOKUP table of the spreadsheet in the table_array input box. It is not necessary to include the title row in this reference; A66:D75 will suffice. The col_index_num is the column of the VLOOKUP table for which the match to the index is desired. For [HCl] this is the second column; enter **2.** The final input to the VLOOKUP function, the range_lookup, is not necessary and can be left blank. Select Finish and press RETURN for the VLOOKUP to be evaluated. Since the cell links of the drop-down boxes G66:G68 are blank, as the dialog sheet has not been run yet, the results of

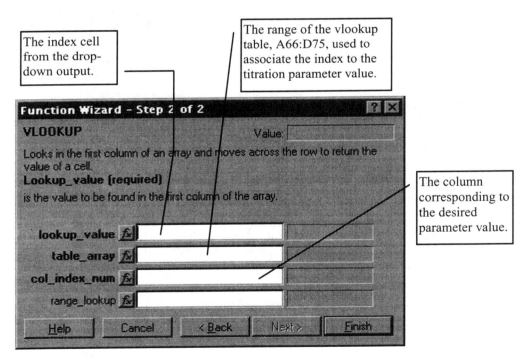

The index cell from the drop-down output.

The range of the vlookup table, A66:D75, used to associate the index to the titration parameter value.

The column corresponding to the desired parameter value.

Figure 8.7. The inputs to the VLOOKUP function for relating the drop-down index to the titration parameter value.

the formulas have all gone to **#N/A.** Enter the value of **1** in each cell of G66:G68 so there is data for the formulas to work on.

- Repeat the steps for the VLOOKUP functions for **[NaOH]** and **Vol of Titrant** in B4 and D3, respectively, changing the lookup_value and col_index_num inputs accordingly.

VBA Code for Running the Dialog Sheet

The code for activating the dialog sheet and hence running the dialog box is simple. All that is needed is a statement to *show* the dialog sheet.

- Select the sheet tab to the right of the dialog sheet and right mouse click to bring up the shortcut menu. Insert a module sheet.
- Enter a few comment lines describing the purpose of the macro.
- Type in the **Sub** line, giving the macro a name.
- Enter the code **DialogSheets("Dialog1").Show** on the line following the Sub line. Follow this by the **End Sub** line. This macro is shown in Table 8.1. Dialog1 refers to the name of the dialog sheet. If you have renamed the dialog sheet, this name needs to be reflected in the above code. There is only one line of code required to activate the dialog sheet, *DialogSheets("Dialog1").Show.* The dialog sheet is ready to run.

The task for the user input in changing the parameters of the strong acid–strong base titration is complete. The input consists of a few Excel sheets: a worksheet, a dialog sheet, and a module sheet. The worksheet contains the named variables and the calculation section for the titration curve, the input ranges and cell links for the drop-down boxes of the dialog sheet, and the graph where the responses to the new parameters are shown. The dialog sheet holds the dialog box and its objects for the user to select new titration parameters. It is linked to the worksheet through formatting the objects of the dialog box. The module sheet contains the code to initiate the dialog box. These all interact when the data sheet is active and the macro is run.

Table 8.1. VBA Vode for Displaying and Running the Dialog Sheet

```
'Macro to run the Dialog Sheet
'
Sub AlterViaDialogBox()
DialogSheets("Dialog1").Show
End Sub
```

Activating the Dialog Sheet with the Macro

- Select the data sheet.
- Display the Drawing toolbar by selecting the drawing tool on the Standard toolbar.
- Choose the Create Button tool and draw a small box just to the left of the graph. When the mouse is released, the Assign Macro dialog box appears. Select the macro for showing the dialog box and choose OK. Highlight the text of the button and give it a descriptive name. If this button needs to be moved about the spreadsheet, it must be selected with the CTRL key held down. This selects it as an object rather than as a control button for activating the macro.
- From the zoom control drop-down box, reduce the display of the spreadsheet if necessary to allow viewing of the entire graph.
- Initiate the dialog box by pressing the macro button. Drag the dialog box by the title bar and move to the left of the sheet in order to view it and the graph simultaneously.
- Select the drop-down boxes for the titration parameters and change their entries. Notice that the graph is automatically updated, displaying the result of the changes. The selections from the drop-down boxes can be changed as often as desired. Selecting OK will close the dialog box.

An alternative way of changing the parameters for the titration is to have a macro that displays input boxes to assign values to the parameters. This was shown in Chapter 3.

Case Study 8.2. Creating a Macro to Change the Titration Parameters through Input Boxes

- Make a copy of the data sheet by holding down the CTRL key and dragging the Data Sheet tab to the right. Rename this sheet *Data2*.
- Insert a module sheet next to *Data2*. Enter the code shown in Table 8.2.
- Change the name of the macro.
- Move to the *Data2* sheet tab and select the macro button for editing the name of the button and reassigning the above macro to it. The CTRL key needs to be held down while selecting. Highlight the text and change the name. While the button is selected as an object, place the mouse over it, right mouse click, and choose **Assign Macro.** Select **AlterViaInputBoxes** or the new name, if you have changed it. Choose OK.
- Select outside the macro button to activate it as an event object and rese-

Table 8.2. The VBA Code to Guide User Input in Changing the Parameters of the Titration through Use of Input Boxes

```
'Macro to ask user to change any of the initializing parameters
'through input boxes

Sub AlterViaInputBoxes()
10 choice = InputBox(prompt:="Enter the parameter you would like to change.  Your choices are _
   HCl, NaOH or Aliquot.", Title:="Titration Parameter")
   Select Case choice
   Case Is = "HCl", "hcl", "Hcl", "HCL", "h"
   newhcl = InputBox(prompt:="New HCl concentration, in molarity", Title:="New [HCl]")
   Range("b3").Formula = newhcl
   Case Is = "NaOH", "NAOH", "naoh", "n"
   newnaoh = InputBox(prompt:="New NaOH concentration, in molarity", Title:="New [NaOH]")
   Range("b4").Formula = newnaoh
   Case Is = "Aliquot", "aliquot", "a"
   newaliquot = InputBox(prompt:="New Aliquot volume, in milliliters (mL)", Title:="New _
   Aliquot Volume")
   Range("d3").Formula = newaliquot
   Case Else
   GoTo 10
   End Select
GoAgain = InputBox(prompt:="Would you like to change another parameter?  Enter y or n", _
Title:="Repeat Choices Message")
   Select Case GoAgain
   Case Is = "y", "Y"
   GoTo 10
   Case Is = "n", "N"
   End
   Case Else
   End
   End Select
End Sub
```

lect it to initiate the macro. Follow the instructions of the input boxes to make changes to the titration parameters.

These two macros are examples of how VBA objects and code can be added to enhance the operation of a spreadsheet for aiding user interaction. With a dialog box, the user can easily visualize the effects of changes made to the titration parameters. This should demonstrate how easy it is to enhance a spreadsheet with VBA. It is hoped that this case study will encourage you to create your own VBA procedures and that you will continue to use VBA in your spreadsheets.

8.2. BUFFER SOLUTIONS

The preparation of buffer solutions are an important part of analytical chemistry. One aspect of determining a buffer solution is to calculate the ratio of the concentrations of the acid (or base) and its conjugate pair. With this ratio, the buffer solution can be prepared for the desired pH. The following case study illustrates how a spreadsheet can be designed to allow a user to calculate this ratio for a particular pH from a list of buffer pairs.

With HA representing an acid and A^- as the conjugate base, the buffer pair concentration ratio is found from rearranging the equation for the dissociation of an acid,

$$\frac{[HA]}{[A^-]} = \frac{[H^+]}{k_a}. \tag{8.1}$$

A spreadsheet in conjugation with a dialog sheet can be designed to calculate the concentration ratio for a chosen buffer pair. Drop-down and text boxes can be used to guide user selections of buffer pair and desired pH for the calculation. This in turn changes the entries in the calculation section of the spreadsheet.

Case Study 8.3. Designing a Spreadsheet to Calculate the Concentration Ratio of a Selected Buffer Pair at a Particular pH

A dialog sheet is used to hold drop-down boxes used for choosing the buffer pairs and pH. The result of the spreadsheet calculation is echoed in a text box on the dialog.

Preparing the Spreadsheet

- Select the next sheet tab and rename to *buffers* by double clicking the sheet tab.
- Enter the data shown in Table 8.3 starting in cell A2. This area lists the buffer pairs with their K_a along with an indexing column for use with the drop-down box of the dialog sheet.
- The next area of the spreadsheet is the table for the pH entries to its drop-down box in the form of a VLOOKUP table. Type in the titles **Index** and **pH** in D2:E2. Enter the data for these columns through the Edit_Fill_

Table 8.3. The Partial Spreadsheet Showing the VLOOKUP Table of the Buffer Pairs, Their K_a and Index[1]

	A	B	C
1			
2	Buffer Pairs	Index	ka
3	HOAc-OAC$^-$	1	1.80E-05
4	NH_4Cl-NH_3	2	5.56E-10
5	H_2PO4^- - $HPO4^{2-}$	3	6.20E-08
6	H_3BO_3 - $H_2BO_3^-$	4	7.30E-10
7	HCO_3^- - CO_2^{2-}	5	4.20E-07
8	HOCN - OCN$^-$	6	1.20E-04

[1]This will be used by the buffer pair dialog box and VLOOKUP function.

Table 8.4. The Calculation Section of the Buffer Spreadsheet (Contains Cell Links, Vlookup Functions, and Formulas)

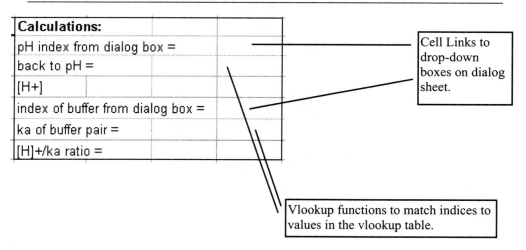

Calculations:

pH index from dialog box =

back to pH =

[H+]

index of buffer from dialog box =

ka of buffer pair =

[H]+/ka ratio =

Cell Links to drop-down boxes on dialog sheet.

Vlookup functions to match indices to values in the vlookup table.

Series menu. Increment the Index from 1 to 70 by 1. Range the *p*H from 0.4 to 14.0, incrementing by 0.2.

- The calculation section of the spreadsheet defines those cells that will serve as links to the objects of the dialog box. Starting in F2, type in the entries of Table 8.4. The cell links for the drop-down boxes are I3 and I6 for the *p*H and buffer pair, respectively. These will be assigned in the next step. VLOOKUP functions will be entered in I4 and I7 to match the index to the *p*H value and the K_a of the selected buffer pair, respectively. Calculation of the [H$^+$] takes place in I5 and that of the buffer pair concentration ratio in I8.

Designing the Dialog Box and Formatting Its Objects

- Insert a dialog sheet. Add objects as shown in Figure 8.8: drop-down objects for the buffer pair and *p*H, edit box for the result of the spreadsheet calculation, labels, and a Calculate button. Select and delete the OK but-

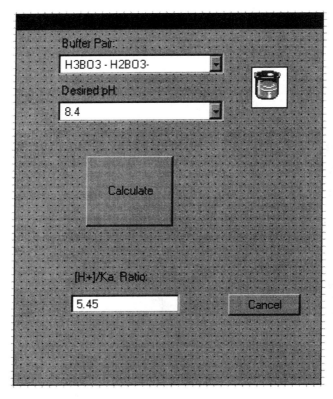

Figure 8.8. The dialog sheet for the buffer concentration ratio spreadsheet.

ton. The Calculate button will serve this purpose. Design the layout to your preference using Figure 8.8 as a guide.

- Format each drop-down object with the appropriate cells as defined on the spreadsheet. Activate the dialog sheet tab, select the buffer pair drop-down, and press CTRL + 1. Click the mouse in the Input Range box then move to the *buffers* sheet and select the range of buffer names. Tab to the Cell Link box, move back to the *buffers* sheet, and select I6. The complete reference to these cells could have been typed in; however, to avoid syntax errors it is best to let Excel enter the reference following the entry manner described above. Edit the Drop Down Lines to **5**.

- Repeat these steps for the *p*H drop-down to assign the corresponding cells of the spreadsheet to this object. Edit the Drop-Down Lines to **10**. This reduces the amount of scrolling necessary.

- Enter the VLOOKUP function for matching the index to the *p*H value in cell I4. Select the cell and choose the Function Wizard or type in **=vlookup(** then press CTRL + A. This is the shortcut to the Function Wizard. The lookup_value is **I3,** the table_array is **D2:E72** and the col_index_num is **2.**

- Do the same for a VLOOKUP in I7 for the buffer pair. This function should read **=VLOOKUP(I6,B3:C8,2).**

- Calculate the [H$^+$] from the *p*H in I5 with **=10^-I4.**

- Enter the formula to divide the [H$^+$] by the K_a in I8. Enclose this with a Round function of 2 decimal places. Hence, **=ROUND(I5/I7,2).**

- Activate the dialog sheet and select the Calculate button. Press CTRL + 1 and select the Control tab. Check the Default and Dismiss boxes. This affords control to the Calculate button.

- Select the Cancel button and rename to Exit Dialog.

The VBA Code and Activating the Macro

- Assigning the results of the calculation on the spreadsheet to the edit box of the dialog sheet needs to be done in VBA code. Insert a module sheet. Enter the code shown in Table 8.5.

The first line in the code is a *Dim* statement for declaring the ratio variable as double precision since the variable is the calculation result. The next line minimizes the *buffers* spreadsheet for visual effect. Hence, only the dialog box will appear during running. The show line displays the dialog box. The next line, *ratio = Range("I8").Value,* assigns the value of I8 to the ratio variable. The purpose of the *With* statement is to assign this variable to the Edit box on

Table 8.5. The VBA Code for Running the Dialog for the *buffer* Spreadsheet.

```
   DialogSheets("BufferDialog").Show
   ratio = Range("I8").Value
      With DialogSheets("BufferDialog")
         .EditBoxes(1).Text = ratio
      End With
 ' displaying an input box to prompt the user to redisplay the dialog sheet
      GoAgain = InputBox(prompt:="Calculate for another buffer pair? Enter y or n.", _
      Title:="Repeat Dialog Message")
      Select Case GoAgain
      Case Is = "y", "Y"
      GoTo 88
      Case Is = "n", "N"
      GoTo 99
      Case Else
      GoTo 99
      End Select
 99 ActiveWindow.WindowState = xlMaximized
End Sub
```

the dialog sheet. Notice that the sequence of lines, *show,* ratio assignment, and *With,* are repeated. The first occurrence displays the dialog sheet with the existing result in I8 echoed in the text box. In the second instance, invoked by pressing the Calculate button, the user-selected changes to the buffer pair and *p*H drop-down boxes calculate a new ratio, and this is displayed in the text box. The Calculate button does not actually perform the ratio calculation; this is done immediately on the spreadsheet when changes have been made to the buffer pair and *p*H drop-down boxes. Pressing this button refreshes the dialog, specifically the edit box, reflecting the change in I8 from when the dialog was first displayed. This button could have been named Refresh Dialog. After performing a calculation on the dialog, the user presses the Exit button and the *InputBox* line of the macro is executed. This displays a message box as laid out in this line. The next sequence of lines are for the *Select Case* that controls the choices the user may make in the message box. The last line maximizes the *buffers* sheet and the macro finishes. The complete *buffer* spreadsheet is shown in Figure 8.9.

Further modifications to the dialog/spreadsheet can be made. For instance, add another buffer pair. This will require adjusting its drop-down range on the dialog sheet. Drawing objects can be inserted onto the dialog box to dress it up. With this working macro, try implementing modifications to the code. For example, experiment with the other VBA control objects of the Forms toolbar.

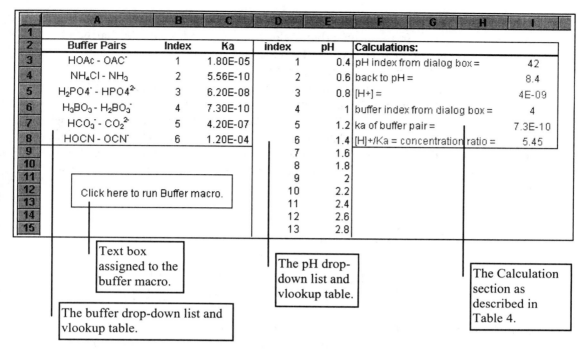

	A	B	C	D	E	F	G	H	I
1									
2	Buffer Pairs	Index	Ka	index	pH	Calculations:			
3	HOAc - OAc⁻	1	1.80E-05	1	0.4	pH index from dialog box =			42
4	NH₄Cl - NH₃	2	5.56E-10	2	0.6	back to pH =			8.4
5	H₂PO4⁻ - HPO4²⁻	3	6.20E-08	3	0.8	[H+] =			4E-09
6	H₃BO₃ - H₂BO₃⁻	4	7.30E-10	4	1	buffer index from dialog box =			4
7	HCO₃⁻ - CO₂²⁻	5	4.20E-07	5	1.2	ka of buffer pair =			7.3E-10
8	HOCN - OCN⁻	6	1.20E-04	6	1.4	[H]+/Ka = concentration ratio =			5.45
9				7	1.6				
10				8	1.8				
11				9	2				
12	Click here to run Buffer macro.			10	2.2				
13				11	2.4				
14				12	2.6				
15				13	2.8				

Text box assigned to the buffer macro.

The pH drop-down list and vlookup table.

The Calculation section as described in Table 4.

The buffer drop-down list and vlookup table.

Figure 8.9. The *buffer* spreadsheet.

Spend time in this way for all the macros written or recorded in Chapter 3 and this chapter. Playing with working code is a very useful method for advancing one's knowledge of a programming language. Always ensure a backup copy of the working macro has been saved before embarking on modifications. Holding down CTRL and dragging the module sheet tab to the right or left is a quick way to create a copy of a sheet.

8.3. THE MARCUS TREATMENT OF ELECTROCHEMICAL KINETICS

8.3.1. Surface Immobilized Redox Couples

The Marcus theory considers a reaction surface to be parabolic, that is, potential energy varies parabolically along a reaction coordinate, rather than linearly as assumed in the traditional Butler–Volmer approach to electrode kinetics. Moreover, the Marcus theory considers contributions to the rate from states in the electrode at potentials other than the Fermi level.

Several models based on these concepts have been developed to explain the potential dependence of the heterogeneous electron transfer rate, that is, the rate of electron transfer from an underlying electrode to a redox couple that is adsorbed on its surface or from the couple to the electrode surface. Our example is for a situation where there is weak electronic communication between the electrode and the adsorbed species. This situation is typically encountered when the active site for electron transfer is located at a relatively large distance from the electrode surface, greater than 20 Å. Under these circumstances, the following expressions can be used to calculate rate constants for reduction, k_{Red} (addition of an electrode to the adsorbate), and oxidation, K_{Ox} (removal of an electron from the adsorbate), as a function of the overpotential (free energy) for electron transfer:

$$k_{Red}(\eta) = A_{et} \int_{-\infty}^{\infty} \exp\left[-\frac{(\epsilon_F - \epsilon + \eta + \lambda)^2}{4\lambda kT}\right]\left(\frac{1}{1 + \exp[(\epsilon - \epsilon_F)/kT]}\right) d\epsilon, \quad (8.2)$$

$$k_{Ox}(\eta) = A_{et} \int_{-\infty}^{\infty} \exp\left[-\frac{(\epsilon_F - \epsilon + \eta - \lambda)^2}{4\lambda kT}\right]\left(\frac{\exp[(\epsilon - \epsilon_F)/kT]}{1 + \exp[(\epsilon - \epsilon_F)/kT]}\right) d\epsilon. \quad (8.3)$$

ϵ_F is the Fermi level of the electrode (the applied potential), ϵ is the energy of a given state in the electrode, λ is the reorganization energy, η is the overpotential (the applied potential relative to the formal potential), k is the Boltzmann constant

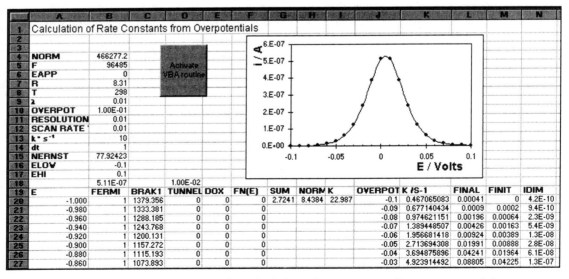

Figure 8.10. The spreadsheet for calculation of rate constants from overpotentials.

$(1.38\times10^{-23}$ J K$^{-1})$ and T is the absolute temperature (298 K $= 25°C$, room temperature).

Case Study 8.4. Development of a Spreadsheet to Vary the Input Variable to a Complex Calculation

The following Excel spreadsheet (Figure 8.10) calculates k_{Ox} by numerically evaluating the integrals in equation 8.2. Input parameters, cells B13 and B8, re-

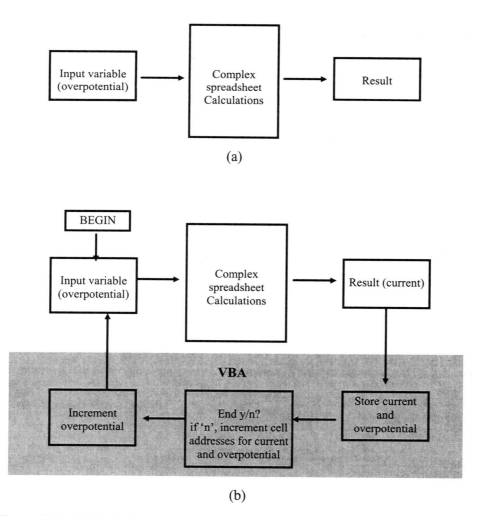

Figure 8.11. (a) Limitations of normal spreadsheet calculation. (b) External control of spreadsheet variables and their location by VBA provides additional flexibility.

Table 8.6. VBA Code for Calculation of Overpotential for the Marcus Plot

```
'VBA Code for Marcus Plot
Sub RFkinetics()
RESOL = Cells(11, 2)
ELOW = Cells(16, 2)
EHIGH = Cells(17, 2)
Index = ELOW * 10 + (20 + Abs(ELOW * 10))
Worksheets("Marcus").Range("j20:k550").Clear
'Normalisation of the heterogeneous electron transfer rate constant
OVERPOTENTIAL = 0
Cells(10, 2).Value = OVERPOTENTIAL
Cells(20, 8).Value = Cells(13, 2).Value / Cells(20, 7).Value
'Calculating the potential dependence of the rate constant
For OVERPOTENTIAL = ELOW To EHIGH Step RESOL
Cells(10, 2).Value = OVERPOTENTIAL
Cells(Index, 10).Value = OVERPOTENTIAL
rate_constant = Cells(20, 9).Value
'Cells(Index, 11).Value = Log(rate_constant) / Log(10) 'log is natural log
Cells(Index, 11).Value = rate_constant
Index = Index + 1
Next OVERPOTENTIAL
Max = -1E+37: Min = 1E+37
For n = 20 To 500
   If Cells(n, 14) > Max Then Max = Cells(n, 14): Index = n
Next n
Cells(18, 2) = Cells(Index, 14)
Cells(18, 4) = Cells(Index, 10)
End Sub
```

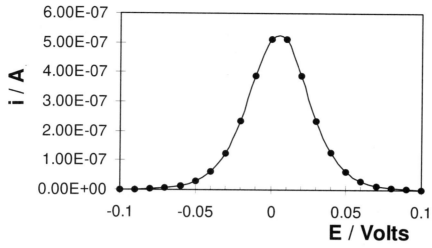

Figure 8.12. Automatically generated i/E Marcus plot using the worksheet *marcus*.

spectively, are the standard heterogeneous electron transfer rate constant k^o and the reorganization energy λ. The rate constants are calculated every 4 mV, and the integrations are performed over the range plus or minus 9 V about the Fermi level at a resolution of 50 mV. The user can change these parameters within the spreadsheet and see that small changes in the integration resolution have little impact on the final curves. The prefactor, A_{et}, is evaluated by first performing the integration at zero overpotential where $k_{Ox} = k_{Red} = k^o$. This prefactor is then used to calculate rate constants at any desired overpotential.

The problem with a standard spreadsheet approach to the above task is that the goal is to evaluate the effect of overpotential on the heterogeneous rate constant and from the latter to calculate the electrode current. The integral in equations (8.2) and (8.3) is evaluated by summing the function over the range of interest. This is achieved in cell G20, which contains the integral formula **=SUM(F20:F168).** Consequently, it is impossible to vary the overpotential in the normal spreadsheet manner, as only one value for the overpotential can be fed into the summation range of cells at a time. However, using VBA it is possible to vary the value in the cell holding the overpotential (cell B10) and to capture the result of the calculation (rate constant, cell I20). This value can then be transferred to another location in parallel with the overpotential value used to generate it, and the cell addresses can be incremented for the next run through the loop (overpotentials stored sequentially in cells J20:J40, equivalent rate constants in cells K20:K40). The VBA routine thus allows the value of a single cell, which determines the output of an entire spreadsheet calculation, to be externally controlled and the results stored in "safe" locations. The concept is summarized schematically in Figure 8.11 (see p. 233), and the VBA code is listed in Table 8.6 (see p. 234). Linking the array of overpotentials and rate constants to a chart window gives an automatic graphical response as each point on the i/E curve is calculated. A sample output graph is illustrated in Figure 8.12.

APPENDIX A

ROUNDING ERRORS IN EXCEL

Users should be careful about using coefficients returned by the built-in curve fitting options available under Excel because of the potential for large errors generated by the default rounding of coefficients in certain cases. This is particularly so for polynomial fits involving higher orders, where even apparently small rounding effects are magnified to huge errors in predicted values by the model. Coefficients returned by Excel should always be examined in a critical manner, and verified before using in a predictive manner. This is easily achieved by calculating the predicted y-values of the original x-data array and comparing to the original y-values. As an example, let us model the data in Table A1 using a 4th order polynomial fit.

The procedure is as follows:

- enter the data into Excel along with the X and Y identifiers;
- Select the two data arrays;
- Plot the data using the chart wizard (scattergraph without line option);
- activate the chart by double-clicking the mouse left-hand button (LHB);
- select the data (single click LHB);
- Select *insert_trendline* from the menubar;
- Choose the *polynomial* option and set to *4th order;*
- From the options sub-menu, check the *display equation on chart* option.

The result is a reasonably good fit to the data (Figure A1).

Table A.1

X	Y
100	215
94.4	215
89.9	210
83.3	185
77.8	160
72.2	130
66.7	105
61.1	83
55.6	65
50	48
44.4	32
38.9	21
33.3	13
27.8	7.9

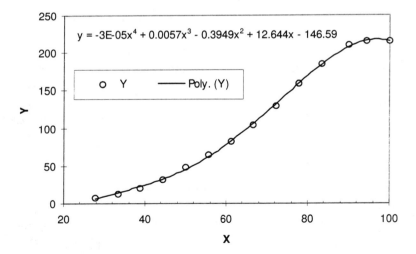

Figure A1. Fit obtained with 4th Order Polynomial to the data in Table A1.

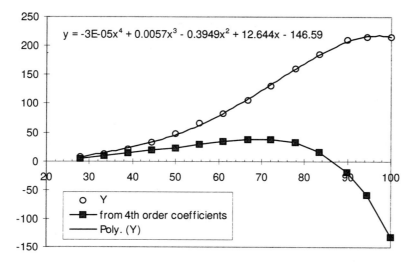

Figure A2. Inaccurate result obtained with the default polynomial coefficients.

However, if we now recalculate the Y-values according to the coefficients in the best-fit equation, the predicted Y-values veer strikingly away from the original Y-values at higher values of X (Figure A2). This suggests a problem with the higher order coefficients of the equation.

The problem is mainly caused by the value of 3×10^{-5} returned for the the 4th order coefficient. This can be corrected by increasing the precision (more significant figures) of the coefficients using the following procedure;

- select the equation in the chart;
- select format from the menu bar (or right click the mouse) and choose *selected data labels;*
- select *number* and use the *scientific* option;
- the number of significant figures can be varied in the *decimal places* box.

Figure A3 shows the much improved fit obtained with the coefficients returned using the scientific number format and *four decimal places.* While this may seem a trivial problem, it is unfortunately the case that students will use coefficients obtained from curve-fitting packages in an uncritical manner. This simple exercise demonstrates the necessity of checking the accuracy of predictions obtained with such models before applying them in real situations.

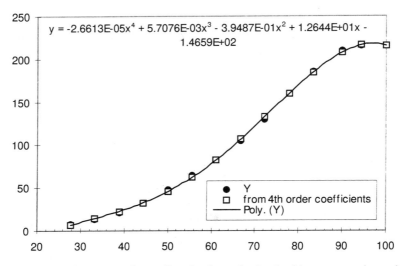

Figure A3. Improved accuracy in predicted values obtained with more precise polynomial coefficients.

INDEX

A1 style, 8
Activity, 114
Analytical Chemistry, 114, 129, 134, 135, 137, 214, 226
Anova, 44
Arrhenius equation, 117

Buffer solutions, 226

Cell reference, 8–9
Chart wizard, 21–25
Chemistry
   Analytical, 114, 129, 134, 135, 137, 214, 226
   Physical, 84, 89, 113, 114, 117, 119, 122, 124, 231
   Quantum, 97, 101, 105, 107
Chromatography peaks, 183
Crystallographic angles, 89, 91
Curve fitting, See also Solver, 32, 47, 51, 53, 45, 61
Custom Charts, 34–35

Data Analysis Toolpak, 42, 44, 47, 51
Data processing, 147
Deconvoluting overlapping peaks, 181
Dialog sheet, 218, 223, 228
Differentiation,
   of titration curves, 134, 141, 143, 144
   for feature enhancement / removal, 163, 166
Digital averaging techniques, 155, 165
Dissociation
   of phosphoric acid, 140
   of weak acid, 130, 132

Edit Fill Series, 15, 27, 71, 103, 108
Editing
   Data, 12–14, 93
   Graphs, see Graphs, Editing graphs,
   Trendline, 53
Enzyme kinetics, 207
Equilibrium,
   Acid-base, 214
   Metal-complex, 124
Electron density plots, 108, 110

Error, 20, 52, 56, 59, 186
　　Relative, 55
Error bars, 57–63
Exponential model, 61–65
Exponentially modified Gaussian model,
　　184, 188

F-test, 41, 42
Flow-injection analysis, 197
Fluorescence decay, 190
Formats, clearing, 13, 68
Formulas
　　Entering, 10, 14, 50
　　Troubleshooting, 37
Function
　　Complex, 98, 103, 105, 108
　　Confidence, 41
　　If, 20, 216
　　IsError, 20
　　Linest, 48
　　Mean *or average,* 40
　　Nested, 19, 20
　　Round, 17
　　Simple, 15, 27
　　Standard deviation, 40
　　User-defined, 20
　　Vlookup, 218, 220, 222, 226, 227
Function Wizard, 16, 17, 20, 39, 50, 222

Gaussian modelling, 174, 179, 181, 183,
　　184, 185, 188
Gran's Plot, 135
Graph
　　Activated, 30
　　Editing graph, 29, 31, 103, 106, 111
　　Embedded, 26
　　Formatting axes, 31, 103, 104, 109,
　　　111
　　Trendline, 53, 55, 59, 61
　　XY scatter, 23, 103, 108, 110, 216
Graphing
　　a simple function, 27
　　with Chart Wizard, 21–25

Importing text files, 148
InputBox, 90, 95, 224–225
Instrumentation–pc interfacing, 148

Kinetics, 117, 207, 231
　　first and second order reations, 119,
　　　122

Logistic model, 197
Ligand replacement reaction, 203
Lineweaver–Burk plot, 208

Macro
　　Calculation macro, 72, 73, 81
　　Formatting macro, 66, 69, 77, 80
　　Function, 82, 84
　　Input data, 90, 95, 224–225
　　to run a Dialog sheet, 223, 230
　　use with Marcus theory, 236
Macros, 3–1, 3–8, 3–9, 3–11, 3–12
　　Code, 77, 80, 81, 90, 95, 223, 225,
　　　230, 236
　　Control button, 89, 224
　　Editing, 79
　　Global, the *personal.xls,* 85
　　Recording, 66, 72
　　Run, 68, 224
　　Subroutine, 77–80, 223
Marcus theory, 231
Membrane potentials, 61, 197
Metal–ligand complexation, 126
Michaelis–Menten equation, 207
Modeling,
　　Exponential, 61–65
　　Exponentially modified Gaussian,
　　　184, 188
　　Fluorescence decay, 190
　　Nikolskii–Eisenman equation, 201
　　Potential relaxation, 61
　　Sigmoid, 197
Molecular orbitals, 105
Moving average, 156, 159, 165, 166, 169
Multiple cell entries, 11

Name box, 2, 7
Name variable, 199, 202
Insert Name Define, 85, 215
Navigating, around a spreadsheet, 7
Nikolskii–Eisenman model, 201

Overpotential, 241

Paste special, 12, 58
Physical Chemistry, 84, 89, 113, 114,
    117, 124, 231
Potential relaxation, 61
Print Area, 38
Printing options, 38
Probability functions, 98

Quantum chemistry, 97, 101, 105, 107

Radial distribution function, 101
Ranges, 9
Rate constants, 119, 120, 122
R1C1 style, 8
References, 73–77
    Absolute, 8, 72
    Relative, 8, 73
Regression
    Linear, 33, 47, 48, 50, 51,. 53
    Non-linear, 47, 61, 183
        Exponential, 55, 190–197
        Polynomial, 55

Savitzky-Golay filter, 157
Schroedinger equation, 97
Sigmoid model, 197
Sine wave, 35
Solid-state chemistry, 84, 89
Solver, 172, 176
    Parameter dialog box, 176, 193, 199
    Parameter explanations, 177
Spectral data processing, 154

Spreadsheets
    3s orbital calculation, 101
    Anova, 44
    Activity coefficient, 114
    Arrhenius equation, 117
    Buffer pair calculation, 226
    Calculation macro, 80–84
    Comparing first and second order re-
        actions, 122
    Data Analysis Toolpak, 42, 44, 51
    Deconvoluting overlapping peaks,
        181
    Dissociation of ethanoic acid, 132
    Dissociation of phosphoric acid, 140
    Electron density plots, 108, 110
    Error bars, 57
    Feature enhancement and removal us-
        ing differentiation and integra-
        tion, 165, 166
    Feature removal by filtering, 162, 166
    First derivative, 110, 129, 134, 141
    First order rate kinetics, 120
    Formatting macro, 66, 69
    Gran's plot, 135
    Importing text file, 148
    Ligand replacement rate constant de-
        termination, 203
    Linest function, 48
    Lineweaver–Burk plot, 208
    Marcus Theory, 231
    Mean and standard deviation, 40, 41
    Metal–ligand complexation, 126
    Modeling fluorescence decay process-
        es, 190
    Modeling the Nikolskii–Eisenman
        equation, 201
    Modeling potential relaxation, 61
    Molecular orbitals, 105
    Moving average filter, 156, 165, 169
    Non-linear, 61
    Quantum leaks, 98
    Radial distribution function, 101
    Savitzky–Golay filter, 157
    Sine wave, 35
    Solid-state chemistry, 84, 89

Spreadsheets *(continued)*
 Solver, 173, 181, 183, 184, 189,190,
  197, 201, 203, 207
 Strong acid-strong base titration, 214
 t- and F-tests, 41, 42
 Temperature conversion, 15
Standard deviation, 40, 41
Statistical funtions, 39, 41, 44, 48
 Anova, 44
 F-test, 41, 42
 t-test, 41, 42

t-test, 41, 42
Text files *see* Text Import Wizard,
Text Import Wizard, 148–153
Titration curves, 129
 Derivative curves, 134, 141
 Strong acid - strong base, 214

 Weak acid - strong base, 132, 140
  Polybasic acids, 137
Tracing,
 Dependents, 37
 Precedents, 37,
Trendlines, *see* Regression,

VBA, 65, 75, 77–80, 80, 81, 223, 225,
  230, 236
VDialog box, 218, 223, 228
VObject browser, 75, 76

Wave functions, 101, 105

XY Scatter graph, *see* Graph, XY Scatter